Nutrition in Pregnancy and Growth

Bibliotheca Nutritio et Dieta

No. 53

Series Editors *P. Walter,* Basel
J.C. Somogyi, Rüschlikon-Zürich

KARGER

Basel · Freiburg · Paris · London · New York ·
New Delhi · Bangkok · Singapore · Tokyo · Sydney

Nutrition in Pregnancy and Growth

Volume Editors *M. Porrini,* Milan
P. Walter, Basel

47 figures and 32 tables, 1996

KARGER

Basel · Freiburg · Paris · London · New York ·
New Delhi · Bangkok · Singapore · Tokyo · Sydney

Bibliotheca Nutritio et Dieta

Library of Congress Cataloging-in-Publication Data
Nutrition in pregnancy and growth / M. Porrini, P. Walter, volume editors.
(Bibliotheca nutritio et dieta; no. 53)
Includes bibliographical references and index.
1. Children – Nutrition. 2. Children – Growth. 3. Pregnancy – Nutritional aspects. I. Walter, Paul, 1933–. II. Porrini, M. III. Series.
TX341.B5 no. 53
[RJ206]
613.2 s–dc20
[612.6]
ISBN 3–8055–6300–0 (hbk.)

Drug Dosage. The authors and the publisher have exerted every effort to ensure that drug selection and dosage set forth in this text are in accord with current recommendations and practice at the time of publication. However, in view of ongoing research, changes in government regulations, and the constant flow of information relating to drug therapy and drug reactions, the reader is urged to check the package insert for each drug for any change in indications and dosage and for added warnings and precautions. This is particularly important when the recommended agent is a new and/or infrequently employed drug.

Printed in Switzerland on acid-free paper by Thür AG Offsetdruck, Pratteln
ISBN 3–8055–6300–0

Contents

Preface

Nutrition is a complex, dynamic field and new information is being discovered almost daily. Knowledge and application of nutritional principles by health professionals are essential in averting the occurrence of disease, in protecting the general health and well-being of the public, and in preventing further complications of illness.

There is no period in life when nutrition is more important than during growth, both in utero and throughout childhood. For this reason 'Nutrition in pregnancy and growth' was the topic of the 31st Symposium of the European Academy of Nutritional Sciences, which was held in Milan, Italy, on May 31 and June 1–2, 1994.

The aim of the symposium was to review the most recent advances in knowledge about such important, basic questions concerning growth and nutrition as 'What are the nutritional needs during pregnancy and lactation?' 'What is the nutritional status of pregnant women in Europe?' 'What is normal growth?' 'What are the nutritional needs which must be met to support normal growth?' Special emphasis was placed on the practical application of new knowledge.

The symposium brought together contributions by acknowledged experts from different European countries and from the USA, who presented invited lectures, free communications and posters, with the idea of giving both a relevant and practical contribution to colleagues in various disciplines (e.g. clinical science, nutrition, public health) and the opportunity of comparing the nutritional situation of different European countries.

The first part of the symposium dealt with nutrition in pregnancy and lactation. The fetus is totally dependent upon the nutritional status of the mother for its development and, in addition, an optimum diet is important for the mother too. Pregnancy brings about many physiological and psychological changes in the mother's body. An inadequate diet during pregnancy may not allow the mother's body to adapt to these changes, which could result in many problems such as miscarriage, premature baby, low-birth-weight baby, and stillborn infant. Beyond the specific needs of the pregnant and lactating woman, many interesting aspects have been studied on the powerful adaptive mechanisms which help to match the mother's energy and nutrient requirements to the available intake and thus to optimise fetal growth and milk production under a wide range of environmental circumstances.

The second part of the symposium consisted of contributions on the nutritional requirements for growth and on some aspects of the evaluation of the nutritional status in children.

Infancy is the period in which growth occurs at a fantastic rate with the weight usually doubling at 6 months of age and being triple the birth weight by 1 year. Consequently, nutrition is essential for an infant to grow well.

The initial source of nutrients is milk. Many authorities recommend breastfeeding as the best means of feeding the infant. However, studies on lactation and human milk are complicated by the considerable variability in milk composition within a single feed, over the course of a day, between breasts and with the duration of lactation. Further, milk is not a stable homogeneous fluid, but rather a complex mixture of interacting components whose physicochemical properties begin to change as soon as it leaves the breast. An important goal of this symposium was to summarize the current state of knowledge in the field, analyse the effect of human milk on the development of infants and its role in the prevention of many diseases, as well as the nutritional requirements during childhood. Other important issues carefully discussed in the symposium were growth and body composition. It is well known that growth is the key outcome measurement for all clinicians dealing with infants and young children. For many years, in fact, the nutritional requirements of infants have been set by measuring the intake of breastfed infants who were growing normally. Consequently, *reference* and *standard* data to use as a general guide for screening and monitoring nutritional status of growing children were presented, as well as new techniques and biochemical markers to study growth rate.

The editors would like to express their gratitude to all the participants of the symposium for their lively and active contributions to the discussions, to all those whose financial support enabled the realization of the symposium, and to Karger Publishers for the collaboration in the publication of these proceedings.

Paul Walter
Marisa Porrini

Porrini M, Walter P (eds): Nutrition in Pregnancy and Growth.
Bibl Nutr Dieta. Basel, Karger, 1996, No 53, pp 1–10

Reproductive Stresses in Undernourished and Well-Nourished Women

Andrew M. Prentice, Gail R. Goldberg, Sally D. Poppitt

MRC Dunn Clinical Nutrition Centre, Cambridge, UK; and
MRC Dunn Nutrition Group, Keneba, The Gambia, West Africa

Introduction

In the past decade a wealth of new evidence has emerged regarding the metabolic and behavioural strategies which help women to balance their daily energy budgets in the face of the additional stresses of reproduction. The data suggest that there are powerful adaptive mechanisms which help to match a mother's energy requirements to the available intake, and thus to optimise fetal growth and milk production under a wide range of nutritional circumstances. This new evidence will be reviewed here with particular emphasis on pregnancy both because it forms one of the central themes of this symposium and because the plasticity in energy metabolism appears to be more marked in gestation than in lactation.

An understanding of how humans fit into the larger scheme of mammalian reproduction helps to explain why the adaptive processes are so powerful in our own species. We will therefore provide a brief evolutionary and interspecies perspective to human gestation before describing the recent findings.

Human Reproductive Stresses in a Zoological Context

Both pre-natal and post-natal growth rates in humans are extremely slow. This is a key feature of human ontogeny which distinguishes it from most other mammals. The phylogenic origin of this trait is believed to lie in the need to maximise the time available for the growth and organisation of a large

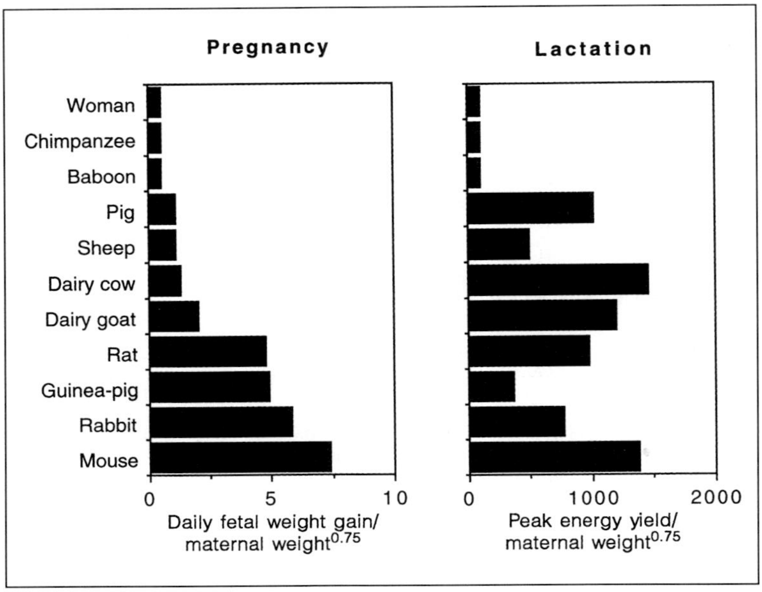

Fig. 1. Estimates of the daily costs of fetal growth and milk synthesis in a variety of mammalian species. Adapted from Prentice and Whitehead [4].

brain [1]. Human gestation is slow even compared to other primates. The expected length of human pregnancy, based on allometric scaling equations for primates, would be about 180 days [2]. In fact, it is 100 days longer. Post-natal growth is similarly retarded compared to other primates in whom full adult size and sexual maturity are reached within a few years of birth [3].

This slow growth has a profound impact on the mother's nutritional requirements since it allows her to spread the costs of fetal growth and milk synthesis over a broad interval thus minimising the nutritional stress when expressed on a daily basis [4]. This is illustrated in figure 1 in which the energetic costs of reproduction, per unit time, are compared across a number of species. Figure 2 provides a measure of the implications of such differences in terms of actual food intake. It shows increments in food intake above the nonpregnant level which are needed to support pregnancy and lactation in a rat dam (with 8 pups), a ewe (with 2 lambs) and a woman (with a single baby). The rat and the sheep must achieve enormous increases in food intake which require substantial hypertrophy of the digestive tract. In comparison the needs of a woman are very modest. The significance of these differences lies in the fact that species with a high reproductive stress have little option other than

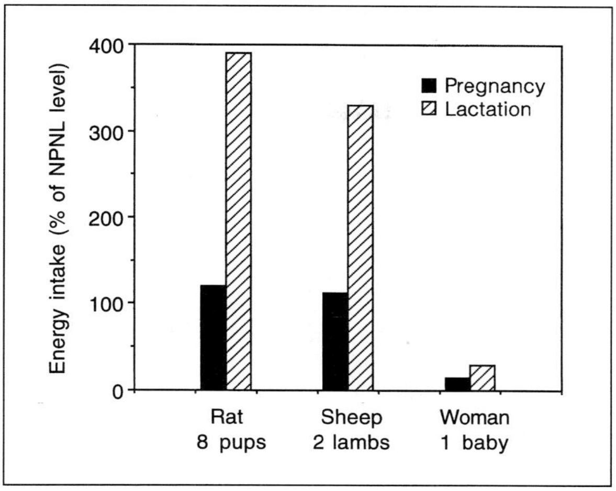

Fig. 2. Incremental energy intake required to support reproduction in rats, sheep and humans. Adapted from Prentice and Whitehead [4]. NPNL = Non-pregnant, non-lactating.

to acquire more food. Failure to find this extra food results in extreme consequences such as abortion, resorption of fetuses, very marked fetal growth retardation, or rejection of the young and infanticide. In humans, however, the marginal daily costs of reproduction are so low that metabolic or behavioural energy-sparing strategies can potentially spare all of the extra energy needed.

It is important, however, to distinguish between the *daily* costs and the *total* costs of reproduction. The total costs for humans are high because, although the offspring only require small amounts of energy and nutrients to deposit new tissue, they have to be maintained for longer. The costs of maintaining the products of human conception are about four times as great as the costs of deposition.

There is a further component to the distinction between humans and most other mammals which may influence energy metabolism. This relates to the size of the maternal fat stores which are available to subsidise reproduction in times of food shortage. Figure 3 shows that even undernourished women have a considerably greater amount of energy stored as fat than most other (non-Arctic and non-aquatic) mammals. Humans as a species tend to be fat [5], and there is a marked sexual dimorphism giving women twice as much fat as men in proportion to total body weight [5]. These large fat stores provide women with yet further flexibility in the strategies that they adopt to meet the energy costs of reproduction.

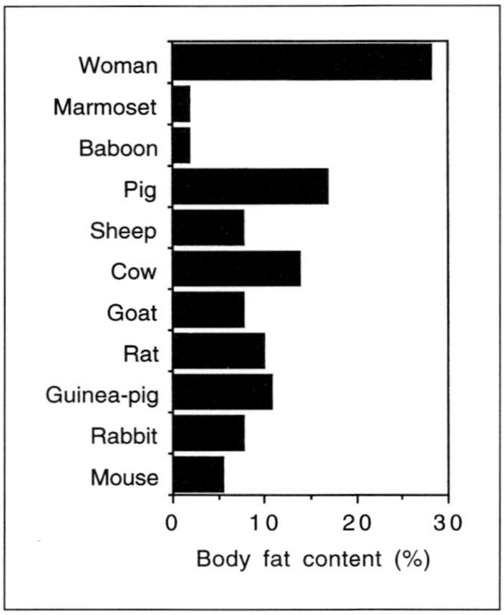

Fig. 3. Approximate estimates of body fat stores in species represented in figure 1.

Evolutionary Influences on Human Reproduction

Propagation of a species can be achieved with a diverse range of maternal investment in the process. At one end of the range are species which spawn millions of eggs, or have frequent and multiple litters, out of which only a small percentage will survive. At the other end is the model currently adopted by many human populations in which the average reproductive pairing produces around two offspring out of which almost 100% survive to reproductive age. The latter strategy involves a much greater maternal investment, and the long reproductive cycle requires the mother to protect the outcome of each conception very carefully if she is to achieve reproductive success over her lifetime. The philogenic shift towards very slow reproduction in women means that it is far better for a mother to make short-term sacrifices to protect her offspring through a period of food shortage than to abort her fetus and initiate a new reproductive cycle when food supplies are more favourable. This is one of the main factors favouring the adoption of efficient energy-sparing strategies in species with slow reproductive cycles.

The metabolic adaptations described below appear to be more powerful in humans than in other mammals studied so far. This can be readily explained by the fact that physiological and behavioural adjustments in energy usage can offset almost all of the reproductive costs in a women (fig. 2), but would make only a trivial impact in most other species and would therefore be unlikely to be amplified by natural selection.

There are strong biological pressures in favour of the natural selection of women who display energy-sparing protective mechanisms in pregnancy. Evolutionary biologists have concluded that seasonal food shortages and famines have played a major part in shaping human metabolism ever since the earliest hominids began to exploit ecological niches which were less favourable than the lush arboreal habitat of their primate ancestors [6, 7]. Those mothers who can sustain a pregnancy with minimal extra nutritional need, or who can carry that pregnancy through a famine, will have the greatest chance of passing on these traits. In our own studies in rural Gambia we have observed that the number of surviving children from women at menopause ranged from one to twelve. Clearly these differences will exert a powerful and rapid selective force in favour of beneficial metabolic traits.

Comparative Studies in Affluent English Women and Poor Gambian Women

For a number of years we have been conducting parallel studies of energy metabolism in affluent, well-nourished women in Cambridge, England, and in poor, under-nourished women in villages surrounding Keneba, The Gambia [9–12]. The Gambian women are subsistence farmers in an area where climatic conditions barely support a single annual crop of staple foods. Their nutritional status is best described as chronically undernourished as judged, for instance, by their average body fat of 20% compared to 30% in the Cambridge women. This chronic undernutrition is exacerbated each year by acute food shortages during a hungry season when the previous year's harvest starts to run out before the current year's crops are ripe. The hungry season causes moderately severe weight loss in all adults, including reproductive women, who use up to 50% of their body fat to buffer the effects of the food shortage [8, 13]. Average birthweight decreases by 200–300 g during the hungry season [14], and milk output also falls slightly although this appears to be more related to a change in maternal feeding patterns than to the mothers' nutritional state [15]. Although this seasonality has a profound impact on energy balance it requires a very detailed analysis which is beyond the scope of this review. We will therefore restrict the presentation of the Gambian data to year-round averages.

Figure 4 illustrates that Gambian women gain only half as much weight during pregnancy as English women. This is perhaps the clearest indicator of the extreme differences in nutritional status between the two populations. Birthweight in the cohort of English women whose metabolic data will be summarised below averaged 3.8 kg, and in the Gambian women averaged 3.0 kg [12].

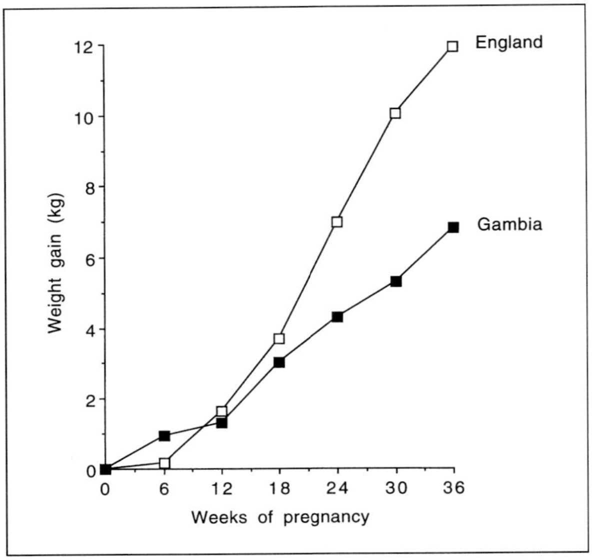

Fig. 4. Comparison of pregnancy weight gain in Gambian and English women. Gambian values represent seasons combined.

In both communities we have conducted prospective, longitudinal measurements of the energy costs of pregnancy starting with baseline data collected in the pre-pregnant state [9–12]. The energy costs of maintaining the products of conception can be estimated from the changes in the mothers' basal metabolic rate (BMR) throughout gestation. These are illustrated in figure 5 for four separate study groups. There is a clear separation between the English women who show an immediate and progressive increase in metabolic rate (as would be predicted on the basis of the accretion of metabolically active tissue), and the Gambian women who show an initial suppression of BMR which then remained close to, or lower than, the pre-pregnant level up to 24 weeks gestation. Gambia I (U) were a group of unsupplemented women studied during years of particular nutritional hardship [9]. The group Gambia I (S) were studied over the same period but were receiving a daily high-energy supplement intended to enhance birthweight [9]. Both of these groups had BMR measured using Douglas bags. Gambia II were a group studied during better years and represent a mixture of supplemented and unsupplemented women [12]. Their BMR was measured by 24-hour whole-body calorimetry which, together with 24-hour calorimetry conducted in the English women [10, 11], demonstrated that the changes in BMR were representative of the overall pattern of metabolism throughout the day.

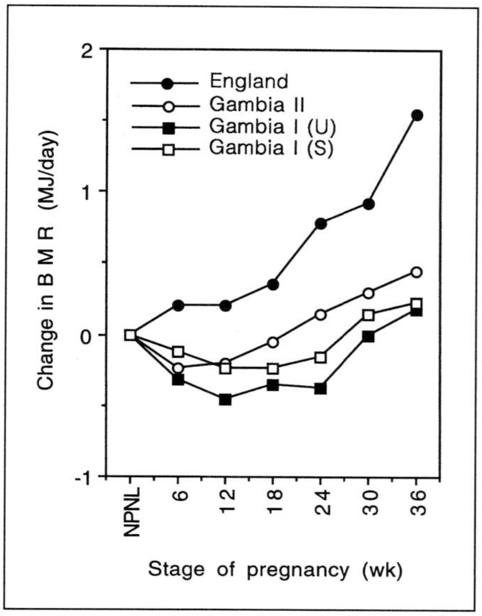

Fig. 5. Longitudinal changes in BMR in Gambian and English women compared to their pre-pregnant baseline. S = Supplemented; U = unsupplemented. [Data from refs 9–12.]

The marked differences between Gambian and English women in their metabolic responses to pregnancy suggest that the change in BMR may in some way be governed by the womens' nutritional status. This impression is emphasised by the fact that the unsupplemented Gambian women show the most energy-sparing response, and that the women studied in the more favourable years showed the least energy-sparing response.

The total maintenance costs of pregnancy can be estimated by integrating the area under the curves in figure 5 and extrapolating to 40 weeks of gestation. These results are shown in figure 6 which reveals a range from 160 MJ in England to –45 MJ in Gambia I (U). Estimates of the energy cost of fat deposition are also shown in figure 6. Once again there is a very wide range from 156 MJ in England to –23 MJ in the unsupplemented Gambian women. Clearly, the marginally nourished Gambian women are managing to achieve a pregnancy with considerably lower metabolic costs than their well-nourished English counterparts.

Other Studies of the Metabolic Costs of Pregnancy from Around the World

Similar longitudinal data have been collected from studies in Sweden, The Netherlands, Scotland, The Philippines and Thailand [16–21]. These results are

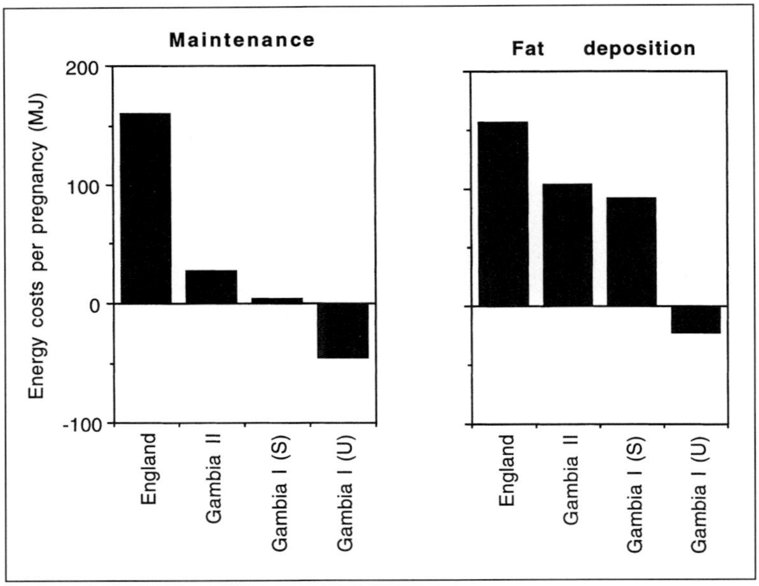

Fig. 6. Cumulative maintenance costs of pregnancy and costs of fat deposition in Gambian and English women.

combined with our own English and Gambian data in figure 7. This illustrates a profound range in the estimated metabolic costs of pregnancy with the highest costs occurring in the most affluent populations. We have shown that the costs of pregnancy between these different groups are closely predicted by each group's pre-pregnant body fat level ($r=0.80$, $p<0.01$) and with pregnancy weight gain ($r=0.94$, $p<0.001$), both of which are themselves correlated [22]. Between-subject analysis has also demonstrated that the energy costs of maintenance are correlated with pre-pregnant body fatness at an individual level; thin women tend to be energy-sparing in pregnancy and fatter women tend to be energy profligate [10, 17].

These results suggest that our Gambian and English data fit into a general pattern in which the energy costs of gestation are modulated in an autoregulatory manner in response to a mother's existing energy status. Those women with abundant energy already deposited as fat or available from the daily diet exhibit energy-profligate increases in metabolism. Those women who are short of energy and have low body reserves exhibit energy-sparing reductions in metabolism. It appears that these may be important mechanisms for protecting fetal growth, since when expressed relative to maternal metabolic

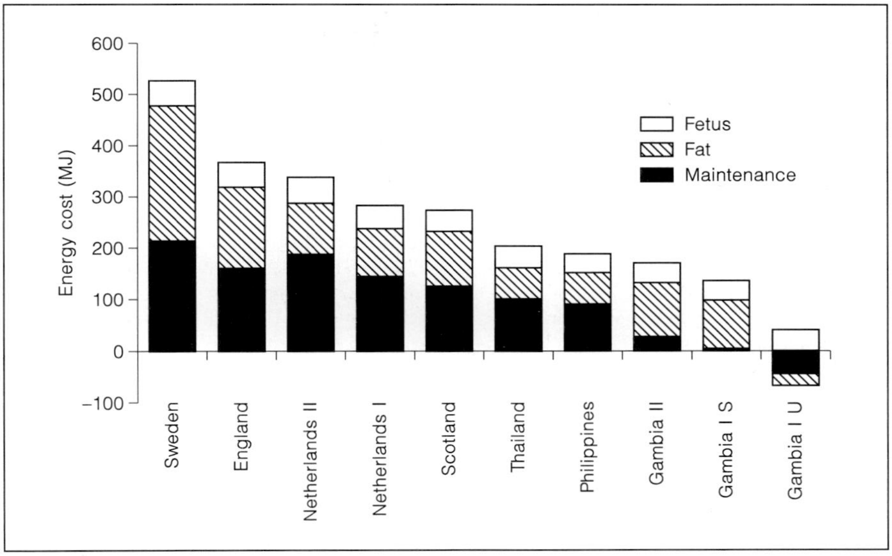

Fig. 7. Total metabolic costs of pregnancy in 10 study groups from 7 countries. Energy deposited in the products of conception was assumed from standard values. Adapted from Poppitt et al. [22].

body size (bodyweight$^{0.75}$), the birthweights in the populations summarised in figure 7 were almost identical (CV = 3%).

Conclusions

We believe that the evidence cited above provides good grounds for believing that human metabolism possesses special plasticity during pregnancy and that there are sound teleological reasons to expect this to be the case. These metabolic adjustments are a powerful means of matching a woman's energy needs to her nutritional circumstances, and almost certainly play a prominent role in maintaining reasonable fetal growth rates in all but the most extreme conditions. The effectiveness of these strategies should not be interpreted as indicating that the provision of adequate nutrition is not an important issue in pregnant women. The protective mechanisms, though impressive in achieving the short-term survival of the fetus, are by no means perfect, and they may have unknown detrimental effects on the longer-term well-being of both the mother and her baby.

References

1 Martin RD: Adaptation and body size in primates. Z Morphol Anthrop 1980;71:115–124.
2 Leutenegger W: Maternal-fetal weight relationships in primates. Folia Primatol 1973;20:280–293.
3 Payne PR, Wheeler EF: Comparative nutrition in pregnancy and lactation. Proc Nutr Soc 1986; 27:129–138.
4 Prentice AM, Whitehead RG: The energetics of human reproduction. Symp Zool Soc Lond 1987; 57:275–304.
5 Pond CM: Adipose tissue in human evolution; in Roede M, Wind J, Patrick J, Reynolds V (eds): The Aquatic Ape: Fact or Fiction? London, Souvenir Press, 1991, pp 193–220.
6 Johnston FE: Seasonality and human biology; in Ulijaszek SJ, Strickland SS (eds): Seasonality and Human Ecology. Cambridge, Cambridge University Press, 1993, pp 5–16.
7 Foley RA: The influence of seasonality on hominid evolution; in Ulijaszek SJ, Strikland SS (eds): Seasonality and Human Ecology. Cambridge, Cambridge University Press, 1993, pp 17–37.
8 Prentice AM, Whitehead RG, Roberts SB, Paul AA: Long-term energy balance in childbearing Gambian women. Am J Clin Nutr 1981;34:2790–2799.
9 Lawrence M, Lawrence F, Coward WA, Cole TJ, Whitehead RG: Energy requirements of pregnancy in the Gambia. Lancet 1987;ii:1072–1076.
10 Prentice AM, Goldberg GR, Davies HL, Murgatroyd PR, Scott W: Energy-sparing adaptations in human pregnancy assessed by whole-body calorimetry. Br J Nutr 1989;62:5–22.
11 Goldberg GR, Prentice AM, Coward WA, Murgatroyd PR, Davies HL, Wensing C, Black AE, Harding M, Sawyer M: Longitudinal assessment of energy expenditure in pregnancy by the doubly-labelled water method. Am J Clin Nutr 1993;57:494–505.
12 Poppitt SD, Prentice AM, Jequier E, Schutz Y, Whitehead RG: Evidence of energy-sparing in Gambian women during pregnancy: A longitudinal study using whole-body calorimetry. Am J Clin Nutr 1993;57:353–364.
13 Prentice AM, Diaz E, Goldberg GR, Jebb SA, Coward WA, Whitehead RG: Famine and refeeding adaptations in energy metabolism; in Anderson GH, Kennedy AH (eds): The Biology of Feast and Famine: Relevance to Eating Disorders. San Diego, Academic Press, 1992, pp 245–267.
14 Prentice AM, Cole TJ, Foord F, Lamb WH, Whitehead RG: Increased birthweight following prenatal dietary supplementation of rural African women. Am J Clin Nutr 1987;46:912–925.
15 Prentice AM, Paul AA, Prentice A, Black AE, Cole TJ, Whitehead RG: Cross-cultural differences in lactational performance; in Hamosh M, Hamosh P (eds): Maternal Environmental Factors in Human Lactation. New York, Plenum Press, 1986, pp 13–44.
16 Forsum E, Sadurskis A, Wager J: Resting metabolic rate and body composition of healthy Swedish women during pregnancy. Am J Clin Nutr 1988;47:942–947.
17 Spaaij CJK: The efficiency of energy metabolism during pregnancy and lactation in well-nourished Dutch women; PhD thesis, University of Wageningen, the Netherlands.
18 van-Raaij JMA, Vermaat-Miedema SH, Schonk CM, Peek MEM, Hautvast JGAJ: Energy requirements of pregnancy in the Netherlands. Lancet 1987;ii:953–955.
19 Durnin JVGA, McKillop FM, Grant S, Fitzgerald G: Energy requirements of pregnancy in Scotland. Lancet 1987;ii:897–900.
20 Thongprasert K, Tanphaichitre V, Valyasevi A, Kittigool J, Durnin JVGA: Energy requirements of pregnancy in rural Thailand. Lancet 1987;ii:1010–1012.
21 Tuazon MA, van-Raaij JMA, Hautvast JGAV, Barba CVC: Energy requirements of pregnancy in The Philippines. Lancet 1987;ii:1129–1130.
22 Poppitt SD, Prentice AM, Goldberg GR, Whitehead RG: Energy-sparing strategies to protect human fetal growth. Am J Obstet Gynecol 1994;171:118–125.

Prof. Dr. A. M. Prentice, Dunn Clinical Nutrition Centre, Hills Road, Cambridge CB2 2DH (UK)

Porrini M, Walter P (eds): Nutrition in Pregnancy and Growth.
Bibl Nutr Dieta. Basel, Karger, 1996, No 53, pp 11–22

Nutritional Studies on Pregnant Women in Umbria (Italy)

Adalberta Alberti-Fidanza[a], *Angela Maurizi Coli*[a], *Irene Ottavi*[b], *Walter Panarelli*[b]

[a] Istituto di Scienza dell'Alimentazione, Università degli Studi,
[b] Divisione di Ostetricia, Presidio Ospedaliero di Gubbio, USLL Alto Chiascio, Perugia, Italy

Introduction

The multiplicity of factors affecting the total nutrition process of pregnant women and fetus creates a number of problems in defining nutritional requirements during pregnancy.

The pregnant woman's capacity of adaption which is particularly efficient, increases notably the difficulty of understanding the nutritional interpretation of the question.

The rate of this adaption depends on the nutrient and, presumably, for the same nutrient, on the stage of pregnancy and on the respective nutritional status of the mother and fetus [1].

Nutritional research on pregnant women is carried out in developing countries to a greater extent than in developed countries because of widespread malnutrition. However in the so-called developed societies over the past few years there has been increasing scientific interest in nutrition questions regarding pregnant women and neonates.

Today in these countries new nutritional behaviour, various environmental factors and the spread of incorrect nutrition information from many sources may all threaten the nutritional status of pregnant and lactating women and infants [2].

Some results obtained from studies carried out in the last few years on groups of pregnant women on a small scale in Assisi (study A) and on a larger scale in Gubbio, Umbertide and Perugia, all areas of the Umbria region, in Central Italy (study B) are presented here [2–7].

These are nutritional integrated studies as they examine a set of variables involved in the same phenomenon, that is the nutritional event during pregnancy. They include measurements of anthropometric variables, food habits, vitamin and mineral nutritional status, blood lipid pattern and physical activity assessment at each trimester during pregnancy.

Mothers' food preferences and aversions, nutritional knowledge, prejudices and beliefs and smoking history were also evaluated.

Multifactorial longitudinal nutritional studies during the entire period of pregnancy are not very common in the literature.

Food Habits

Food habits before and during pregnancy were evaluated using the dietary history method. During pregnancy women were interviewed about their food habits at the end of the 1st, 2nd and 3rd trimester. They were helped to estimate portion sizes with pictures or food models.

In the preliminary study carried out in Assisi (study A) low intake of milk and cheese and a rather high intake of meat were observed in the 20 pregnant women followed during pregnancy. Small quantities of eggs and fish, and about 200 g/day of cereals were consumed. Cakes, desserts, sugar and honey made up about 70% of the total carbohydrate quota. Saccharose represented a rather high percentage of the daily energy intake. The daily energy intake was, on average, about 2,000 kcal. This figure may be considered satisfactory due to the light physical activity observed during the whole pregnancy. However, in some cases there was a markedly low energy intake. Protein and fat as a percentage of energy were high during the whole pregnancy (about 16 and 40%, respectively).

The eating pattern was rather similar during all three trimesters; in the 2nd trimester a higher intake of energy was observed, about 140 kcal more than in the 1st and 3rd trimester, due prevalently to the higher intake of fruit. Calcium and iron intake were always deficient as well as the polyunsaturated/saturated (P/S) fatty acid ratio. Among vitamins only vitamin A intake in the 2nd trimester was below the recommended amounts [2].

In the larger-scale study (study B) of which we present the results of 53 women who were followed in the 1st, 2nd and 3rd trimester similar characteristics of food choices and consequently of diet compositon were observed (tables 1, 2).

The paired t test analysis shows significant differences in the trend of food intake during pregnancy ($^{*}p < 0.1$) only for cereals between the 2nd and 3rd trimesters.

Table 1. Mothers' daily mean and SD of food intakes (g/day) during pregnancy

Food intake	1st trimester (n=53)	2nd trimester (n=53)	3rd trimester (n=53)
Milk	137±194	137±122	169±188
Cheese	54±49	50±46	54±43
Meat	154±66	166±84	162±68
Eggs	17±15	14±13	13±10
Fish	20±15	22±16	21±30
Oils and fats	30±13	32±14	32±17
Cereals	255±120	259±101	224±112*
Legumes	20±32	17±34	20±42
Vegetables, potatoes	276±106	289±129	279±121
Fruit	407±264	449±228	481±257
Cakes, dessert	52±55	58±61	50±46
Sugar, honey	13±15	12±13	11±11

Significance of difference between 3rd and 2nd trimester: *$p<0.1$.

Similarly as regards energy and nutrient intakes significant differences (*$p<0.1$) were observed only for carbohydrates (prevalently due to the intake of fruit) and retinol between the 2nd and 3rd trimesters and 1st and 2nd trimesters, respectively.

The percentage of energy from macronutrients is kept almost at the same level from the 1st to the 3rd trimester. The percentage of energy from proteins and fat, mostly from animal sources, was high. Consequently, the percentage from carbohydrates was rather low.

The distribution of energy from energy macronutrients did not correspond to the values recommended for the general population. However, the problem of the best distribution of energy from protein, fat and CHO during pregnancy is still unresolved. No expert panel has yet produced guidelines for pregnant women's fat intake, except for the percentage of energy from essential fatty acids [8].

The P/S ratio was always very low, 0.2–0.3, and consequently the essential fatty acid intake probably did not meet the recommended level. The mean intake of all vitamins considered was satisfactory, but it was not for calcium and iron which was always too low.

Table 2. Mother's daily mean and SD of energy and nutrient intakes during pregnancy

	1st trimester (n = 53)	2nd trimester (n = 53)	3rd trimester (n = 53)
Energy, kcal	2,210.3 ± 798.1	2,289.5 ± 605.2	2,162.3 ± 575.0
Energy, MJ	9.1 ± 3.2	9.5 ± 2.5	8.9 ± 2.4
Protein, g	86.0 ± 27.8	89.8 ± 25.0	87.5 ± 22.6
Protein % en.	15.8 ± 2.5	15.8 ± 2.5	16.4 ± 2.7
Fat, g	81.9 ± 30.4	84.6 ± 24.3	82.6 ± 22.5
Fat, % en.	33.7 ± 5.3	33.5 ± 5.7	34.8 ± 5.9
SatFA, g	26.7 ± 12.4	27.6 ± 10.3	26.9 ± 9.2
SatFA, % en.	10.9 ± 2.7	10.8 ± 2.9	11.3 ± 3.2
PUFA, g	6.4 ± 2.8	6.6 ± 2.6	6.6 ± 2.9
PUFA, % en.	2.6 ± 0.7	2.6 ± 0.8	2.8 ± 1.0
P/S	0.2 ± 0.1	0.2 ± 0.1	0.3 ± 0.1
CHO, g	284.1 ± 118.0	290.8 ± 91.5	268.2 ± 93.5*
CHO, % en.	47.8 ± 6.3	47.4 ± 6.3	46.0 ± 7.2
Fibre, g	32.2 ± 13.3	32.6 ± 11.3	31.9 ± 12.8
Ca, mg	848.2 ± 439.6	872.4 ± 351.9	905.6 ± 370.7
Fe, mg	13.1 ± 4.9	13.7 ± 4.1	13.2 ± 4.2
Zn, mg	7.6 ± 3.4	7.5 ± 2.5	7.2 ± 2.5
Cu, mg	1.82 ± 0.70	1.92 ± 0.64	1.95 ± 0.77
Thiamin, mg	1.04 ± 0.39	1.06 ± 0.32	1.06 ± 0.32
Riboflavin, mg	1.34 ± 0.64	1.38 ± 0.38	1.48 ± 0.57
Niacin eq., mg	30.5 ± 10.3	31.3 ± 9.5	31.1 ± 9.0
Retinol eq., µg	938.5 ± 633.5	1,276.8 ± 1,072.3	1,321.4 ± 1,417.0
Ascorbic acid, mg	151.5 ± 92.2	154.7 ± 87.6	150.8 ± 73.9

Significance of difference between 3rd and 2nd: $^{*}p < 0.1$.

Anthropometry

In study A the average weight gain observed during the whole pregnancy was 10.4 kg (the prepregnancy BMI was 27.9). Further information on the antropometry can be found in a previous paper [2].

In study B, weight gain during the whole pregnancy was on average 12.5 kg. The weight gain according to BMI met the recommended values for about 50% of the women both with a prepregnant BMI < 19.8 and between 19.8 and 26.0 [9]. The anthropometric values for the group of 53 women are presented in table 3.

Among the skinfolds considered only the thigh skinfold changed significantly during the whole pregnancy, from the 1st to the 3rd trimester

Table 3. Changes in anthropometric variables and hand grip strength during pregnancy (mean ± SD)

	1st trimester (n = 53)	2nd trimester (n = 53)	3rd trimester (n = 53)
Women's characteristics			
Age, years	26.4 ± 4.0	26.7 ± 4.1	27.1 ± 4.0
Pregnancies, n	1.6 ± 0.9	1.6 ± 0.9	1.6 ± 0.9
During pregnancy			
Stage of pregnancy, weeks	10.3 ± 2.3	25.2 ± 0.6	37.0 ± 0.9
Weight, kg	59.5 ± 9.9	65.9 ± 9.9	71.0 ± 9.9
Weight change, kg	1.0 ± 2.4	7.4 ± 3.6	12.5 ± 4.6
Weight change, %	1.8 ± 4.1	13.1 ± 6.7	22.0 ± 8.6
BMI	23.2 ± 4.1	25.6 ± 4.0	27.6 ± 3.9
Circumferences, cm			
Arm	26.4 ± 3.2	27.2 ± 3.1	27.4 ± 2.9
Thigh (max)	54.9 ± 5.0	56.9 ± 4.7	57.5 ± 4.6
Thigh (middle)	51.8 ± 5.1	53.6 ± 4.7	54.6 ± 4.8
Calf	34.3 ± 2.5	35.1 ± 2.6	35.9 ± 2.6
Abdomen	86.4 ± 10.7	96.2 ± 9.2	104.2 ± 8.6
Diameters, cm			
Biacromial	34.0 ± 2.6	34.0 ± 2.6	34.0 ± 2.6
Biiliocristal	28.9 ± 2.0	29.4 ± 1.8	29.6 ± 1.9
Bicondil-femor	9.0 ± 0.7	9.0 ± 0.7	9.0 ± 0.7
Wrist	4.8 ± 0.4	4.8 ± 0.4	4.8 ± 0.4
Handgrip strength (N)			
Right	2.1 ± 0.7	2.1 ± 0.6	2.1 ± 0.7
Left	1.7 ± 0.6	1.8 ± 0.6	1.8 ± 0.6
Skinfolds, mm			
Triceps	17.2 ± 5.9	19.1 ± 5.9	19.3 ± 5.5
Biceps	8.8 ± 4.4	9.6 ± 3.6	9.2 ± 3.3
Subscapular	16.5 ± 8.8	18.8 ± 7.7	19.2 ± 7.2
Suprailiac	12.4 ± 6.7	17.5 ± 6.4	18.2 ± 6.1
Thigh	30.4 ± 7.0	34.3 ± 6.3***	36.0 ± 6.8***
Calf	17.6 ± 5.9	19.2 ± 6.6	19.5 ± 6.6
Thorax	12.8 ± 6.5	14.2 ± 5.9	14.1 ± 5.6
Σ (skinfolds)	114.4 ± 37.6	131.2 ± 33.8	133.5 ± 31.8

(Continued next page)

Table 3 (continued)

	1st trimester (n=53)	2nd trimester (n=53)	3rd trimester (n=53)
Indices			
1	0.95 ± 0.31	1.00 ± 0.30	1.01 ± 0.32
2	0.63 ± 0.18	0.67 ± 0.19	0.69 ± 0.22
3	1.26 ± 0.37	1.33 ± 0.37	1.37 ± 0.44
4	0.38 ± 0.12	0.44 ± 0.11	0.44 ± 0.12
5	0.65 ± 0.20	0.77 ± 0.19	0.79 ± 0.22
Muscle circumferences, cm			
Arm	22.3 ± 2.3	22.7 ± 2.3	23.0 ± 2.1
Thigh	42.2 ± 3.9	42.8 ± 3.9	43.2 ± 3.8
Muscle area, cm^2			
Arm A total	39.8 ± 8.3	41.4 ± 8.4	42.4 ± 7.8
Fat area, cm^2			
Arm B	16.2 ± 7.4	18.3 ± 7.2	18.2 ± 6.5
Arm total A+B	56.0 ± 13.8	59.6 ± 14.0	60.6 ± 12.8

Indices: 1 = Subscapular/triceps; 2 = subscapular/triceps + biceps, 3 = subscapular/(triceps + biceps)/2; 4 = subscapular + suprailiac/triceps + biceps + calf + thigh; 5 = subscapular + suprailiac/triceps + biceps + thigh.

Significance of difference between 2nd and 1st trimester: *** $p < 0.001$; and between 3rd and 2nd trimester: *** $p < 0.001$.

(*** $p < 0.001$). The other six skinfolds and the sum of seven skinfolds showed only a tendency to increase. Their increases except for the biceps skinfold and the increase of the indices were significant only between the 1st and the 2nd trimesters (*** $p < 0.001$) and between the 1st and the 3rd trimesters (*** $p < 0.001$).

The right- and left-hand grip strength never showed significant changes.

All circumferences, except the brachial circumference, increased significantly during the whole pregnancy. For the brachial circumference the increase was not significant only between the 2nd and the 3rd trimesters.

Table 4. Food preferences and aversions during pregnancy

Question	For which of your usual foods did you have a particular craving during pregnancy?	
Answers	Pasta, bread	46%
	Fruit	19%
	Ham and salami, vegetables	6%
	Meat	4%
	Sweets, cheese	3%

Table 5. Nutrition knowledge during pregnancy

Questions	Which foods should be excluded? Why?	
Answers	46% fats	24% they are harmful to health 13% are fattening 63% no specification given
	12% alcohol	70% it is harmful to health 30% no specification given
	7–8% no food, coffee	
	5% bread, pasta	

Food Preferences and Aversions

Generally, when women were interviewed about their preferences and aversions during pregnancy they informed us that they would have liked to eat more pasta and bread or other cereals, but as these foods were considered fattening they unwillingly ate other foods, not realizing that the amount of energy was practically the same.

As we have said already, meat consumption was high, although women did not like it very much during pregnancy, especially in the 1st trimester (table 4).

Food habits before pregnancy were similar to those during pregnancy, but not food preferences.

As food habits may not correspond to the real food preferences, the interpretation of the nutritional needs during pregnancy may be confused.

Nutritional knowledge was poor (table 5). Its source was prevalently popular beliefs, advertising, but, unfortunately, also the medical profession.

Table 6. Changes in blood lipids during pregnancy (mean + SD)

	1st trimester	2nd trimester	3rd trimester
Total Cholesterol, mmol/l	4.47 ± 0.75 (n = 52)	6.47 ± 1.09 (n = 50)	6.91 ± 1.26 (n = 49)
HDL-Chol, mmol/l	1.63 ± 0.50 (n = 51)	2.03 ± 0.59 (n = 45)	1.88 ± 0.53 (n = 47)
HDL-Chol/total cholesterol, mmol/l	0.37 ± 0.11 (n = 51)	0.32 ± 0.09 (n = 45)	0.28 ± 0.08 (n = 47)
Triglycerides, mmol/l	1.07 ± 0.53 (n = 52)	1.86 ± 0.56 (n = 50)	2.83 ± 0.77 (n = 49)

Generally, food habits which were observed certainly cannot be defined as strictly Mediterranean. Alcohol intake was very low.

The energy intake during pregnancy was similar to that observed by other authors in Swedish, French and English pregnant women and also the highest value of energy intake was observed in the 2nd trimester [10].

Smoking Habits

As regards smoking habits of the women examined, generally about 20% did not smoke before or during pregnancy. The smokers cut down drastically. Women who smoked 15–20 cigarettes daily during pregnancy gained only 6 kg from the 1st trimester to the 3rd and their newborns weighed 2,400 g at birth, after a gestation period of 38 weeks [2].

Blood Lipids

The most relevant differences that we observed during pregnancy were for triglycerides (TG) and total cholesterol (T-Chol) (table 6).

TG increased by 74% from the 1st to the 2nd trimester, by 52% from the 2nd to the 3rd and by 164% from the 1st to the 3rd trimester.

T-Chol increased particularly from the 1st to the 2nd trimester. HDL-cholesterol increased from the 1st to the 2nd trimester by 24%, and then decreased slightly.

Blood lipid fractions analysed showed significant variations from the 1st trimester to delivery.

Our results on blood lipid pattern in the course of a normal pregnancy coincided with those obtained by other authors [8–11].

During pregnancy nearly all women develop a trasient hyperlipidemia, but the long-term consequences of these conditions on CHD have not been adequately investigated in controlled clinical studies.

Pregnancy is a transient event in the life of a woman, moreover there is an increased prevalence of angina, cholesterol gallstones, diabetes and obesity in postmenopausal women as a consequence of several pregnancies [8].

Vitamin and Mineral Status

The results of vitamin and mineral nutritional status assessed in both studies (A and B) showed high frequencies of deficiencies, assessed according to the guidelines as proposed by Schrijver [12].

From study A [3], in many cases deficiencies are observed, especially for folic acid, for which about 80% of women showed low blood levels from the beginning until the end of pregnancy.

100% of the women were at risk for any one of the vitamins, 29% at risk for any two, 33% for three and only one person (20%) was at risk for all four of the vitamins: thiamin, riboflavin, folic acid, and retinol.

The results of study B on vitamin and mineral nutritional status are presented in table 7.

Deficiencies are observed particularly for folic acid, piridoxin, ascorbic acid, iron and ferritin. From the paired t test analysis significant differences are observed between the 1st and 2nd trimester and between the 1st and the 3rd trimester for ascorbic acid, Zn, Fe and hemoglobin. Low blood levels of folic acid are observed from the beginning of the pregnancy to the end and the differences are not significant.

As pregnancy advanced not only did the number of deficiencies increase but also the rate of deficiency became more serious.

For iron deficiency the criteria proposed by INACG were used [13]. The results on iron and folic acid deficiency obtained from both studies are rather similar. For study A see the relevant papers [3–7].

In a sub-group of study B, ferritin values <12 µg/l were observed up to the 25th percentile in the 2nd and in the 3rd trimesters.

For folic acid the marginal values (6.8–13.6 nmol/l) were found for 75% of women during the 1st and the 2nd trimester, and up to the 50th percentile in the 3rd trimester. Deficient states were found up to the 25th percentile during the 1st, the 2nd and 3rd trimesters.

Table 7. Vitamin and mineral status during pregnancy

	n	1st trimester mean ± SD	n	2nd trimester mean ± SD	n	3rd trimester mean ± SD
Thiamin, α-ETK TPP effect (ratio)	53	1.1 ± 0.1	53	1.1 ± 0.1	51	1.1 ± 0.1
Riboflavin, α-EGR FAD effect (ratio)	53	1.1 ± 0.1	53	1.1 ± 0.1	51	1.2 ± 0.2
Pyridoxin, α-EGOT PLP effect (ratio)	53	1.6 ± 0.2	53	1.7 ± 0.2	50	1.8 ± 0.4
Folic acid, nmol/l	50	4.3 ± 1.8	49	4.0 ± 2.6	48	4.5 ± 3.5
Vitamin B_{12}, pmol/l	50	472.4 ± 601.1	49	351.7 ± 267.6	48	385.9 ± 407.2
Retinol, μmol/l	13	1.4 ± 0.3	15	1.3 ± 0.4	23	1.1 ± 0.3
β-Carotene, μmol/l	13	0.3 ± 0.2	15	0.4 ± 0.4	23	0.2 ± 0.2
α-Tocopherol, μmol/l	19	15.8 ± 5.9	21	19.4 ± 10.7	32	17.0 ± 10.1
Zinc, μmol/l	50	16.4 ± 3.6	46	12.9 ± 3.5	46	12.6 ± 3.5
Iron, μmol/l	52	17.6 ± 5.6	53	13.7 ± 5.1	48	12.4 ± 4.8
% saturation	50	29.7 ± 10.4	50	18.0 ± 8.2	48	14.8 ± 7.5
Ferritin, μg/l	52	66.1 ± 49.4	51	21.2 ± 19.2	51	14.1 ± 8.6
Hemoglobin, mmol/l	52	8.0 ± 0.6	51	7.4 ± 0.6	48	7.4 ± 0.7
Ascorbic acid, μmol/l	51	22.7 ± 14.2	53	18.0 ± 10.8	50	16.7 ± 11.6

The percentages of inadequacy were low for all four variables (ferritin, transferrin saturation percent, hemoglobin and folic acid) in the 2nd (4%) and 3rd trimesters (6%). The highest percentages of inadequacy for associated variables were found for ferritin + transferrin saturation percent in the 2nd (21%) and 3rd trimesters (24%). The transferrin saturation percent was < 16% up to the 50th percentile in the 3rd trimester and at delivery.

In advancing pregnancy vitamin deficiencies increase considerably.

During the entire pregnancy few women were in a satisfactory status for all nutrients examined.

Among the nutrients considered, significant positive correlation between blood levels and diet intake for vitamin C (* $p < 0.1$) and for iron and hemoglobin (** $p < 0.01$) were observed only in the 2nd and 3rd trimesters.

Data on prepregnancy vitamin and mineral status were not available; pregnant women at risk during pregnancy were probably in unfavourable conditions before pregnancy at least for some nutrients.

These vitamin and mineral deficiencies during pregnancy, although at a subclinical level, occurred in middle class women, despite economic well-being and the great abundance and variety of food which was available.

The decreasing trend in vitamin levels during pregnancy, observed generally by several authors, cannot be explained by the increase alone in circulating plasma volume.

The strong positive correlation observed during pregnancy between the fall in serum vitamin levels into the last weeks and the initial serum levels, has been interpreted as indicative of a 'resetting' of maternal vitamin homeostasis. A redistribution of maternal vitamin stores between the cellular and extracellular compartments has been postulated.

In these and in other studies, except for iron, the decreasing trend of vitamin and mineral levels was in general independent of the respective intake [14].

The nutritional literature gives a number of results from nutritional studies, including nutrient supplementation, carried out on pregnant women in different countries and in different nutritional realities. Particular attention has to be paid when these results are extrapolated, especially from developing to developed countries.

When the diet is extremely poor and the general nutritional status is very unsatisfactory, the capacity of adaption may vary dramatically.

Several authors suggest supplements of vitamins and minerals during pregnancy or before pregnancy in order to begin pregnancy in a satisfactory nutritional status.

Attention has to be given to providing balanced nutrient multi-supplements, especially where minerals are concerned, for example iron and zinc.

If the cutoff points for the assessment of vitamin and mineral states are adequate, deficiencies during pregnancy at a subclinical level and particularly the multideficiencies, probably can have some immediate or future negative effects on mother and newborn, although at a functional level. This is not yet proved as it is very difficult to identify and evaluate.

In conclusion, the complex events of pregnancy, even in the physiological conditions, are an adventure for the mother and fetus and for those scientists who are courageous enough to study them.

References

1 Pitkin RM: Assessment of nutritional status of mother, fetus and newborn. Am J Clin Nutr 1981; 34:658–668.

2 Alberti-Fidanza A, Fidanza R: A nutrition study involving a group of pregnant women in Assisi, Italy. 1. Anthropometry, dietary intake and nutrition knowledge, practices and attitudes. Int J Vitam Nutr Res 1986;56:373–380.

3 Alberti-Fidanza A, Simonetti MS, Mariani Cucchia L: A nutrition study involving a group of pregnant women in Assisi, Italy. 2. Determination of vitamin nutriture. Int J Vitam Nutr Res 1986; 56:381–386.

4 Alberti-Fidanza A, Fidanza R: Pregnancy: A nutritional adventure? In Moyal MF (ed): Dietetics in the 90s. Proc Xth Int Congr Dietetics. London, Libbey, 1988, vol 1, pp 27–31.
5 Alberti-Fidanza A: A nutritional study involving groups of pregnant women in Umbria, Italy: Preliminary results; in Berger E (ed): Vitamins and Minerals in Pregnancy and Lactation. Nestlé Nutrition Workshop series. New York, Raven Press, 1988, vol 16, p 113.
6 Alberti-Fidanza A, Simonetti MS, Coli R, Maurizi A: Nutritional problems of pregnant women in Umbria (Italy); in Lintas C, Spadoni MA (eds): Food Safety and Health Protection. Progetto Finalizzato IPRA. Rome, CNR, 1988, No 28, pp 265–273.
7 Alberti-Fidanza A, Fiorucci G, Simonetti MS: Iron and folates nutriture in pregnant women in Umbria (Italy); in Hercberg S, Galan P, Dupin H (eds): Recent Knowledge on Iron and Folate Deficiencies in the World. Colloque INSERM. Paris, INSERM, 1990, vol 197, pp 115–119.
8 Hachey DL: Benefits and risks of modifying maternal fat intake in pregnancy and lactation [1–4]. Am J Clin Nutr 1994;59(suppl):454S–64S.
9 King JC, Butte NF, Bronstein MN, Kopp LE, Lindquist SA: Energy metabolism during pregnancy: influence of maternal energy status [1,2]. Am J Clin Nutr 1994;59(suppl):439S–445S.
10 King JC, Bronstein MN, Fitch WL, Weininger J: Nutrient utilization during pregnancy; in Bourne GH (ed): Energy Nutrition of Women. Wld Rev Nutr Diet. Basel, Karger, 1987, vol 52, pp 71–142.
11 Hytten FE: Nutrition; in Hytten FE, Chamberlain G (eds): Clinical Physiology in Obstetrics. London, Blackwell, 1980, pp 163–192.
12 Schrijver J: Biochemical markers for micronutrient status and their interpretation; in Pietrik K (ed): Modern Lifestyles, Lower Energy Intake and Micronutrient Status. London, Springer, 1991, pp 55–85.
13 International Nutritional Anemia Consultative Group: Measurements of Iron Status. Washington, Nutrition Foundation Inc, 1985.
14 Van den Berg: Vitamin and mineral status in healthy pregnant women; in Berger E (ed): Vitamins and Minerals in Pregnancy and Lactation. Nestlé Nutrition Workshop Series. New York, Raven Press, 1988, vol 16, pp 93–108.

Prof. Adalberta Alberti-Fidanza, Istituto Scienza dell'Alimentazione, Universita' degli Studi di Perugia, Casella Postale 333, I–06100 Perugia (Italy)

Porrini M, Walter P (eds): Nutrition in Pregnancy and Growth.
Bibl Nutr Dieta. Basel, Karger, 1996, No 53, pp 23–36

Nutrition during Lactation

Margit Hamosh

Division of Developmental Biology and Nutrition, Department of Pediatrics, Georgetown Univesity, Medical Center, Washington, D.C., USA

Introduction

This chapter will discuss the effect of maternal nutrition during lactation as it pertains to the United States. The resurgence of breastfeeding in the 1970s, following its marked decline during the preceding decades [1], has led to renewed interest in the composition of milk and its function in the newborn infant. This concern has resulted in several international meetings and the publication of several books that dealt with establishing the proper techniques for collection and analysis of milk components [2, 3], the effects of maternal and environmental factors on human milk [4], the effect of human milk upon the recipient infant [5, 6], as well as the protection against disease conferred by human milk upon the newborn [7]. The most recent and complete information about the composition of human milk as well as a comparison between the latter and cow's milk appeared in 1995 [8].

The specific topic of nutrition as it pertains to pregnancy and lactation was considered in depth by the National Academy of Sciences which convened a group of experts whose joint work resulted in the publication of two reports: Nutrition during pregnancy [9] and Nutrition during lactation [10].

In this chapter I will summarize the findings published in 'Nutrition during lactation' and will update the information since the publication of this report.

The most important finding of the above report is that 'lactation is robust', i.e. mothers are able to produce milk of sufficient quantity and quality to support growth and promote the health of infants, even when the mother's supply of nutrients and energy is limited.

Population Surveys to Determine Who Is Breastfeeding in the USA

A close examination of the women who breastfeed shows that they tend to be well educated, older and Caucasian. In the USA, 52% of women initiate beastfeeding but only 19.6% breastfeed their infants at 5–6 months of age [10]. However, while white and Hispanic women have similar rates of initiation of breastfeeding (58.5 and 48.4%, respectively) and similar rates of long-term, 5–6 months' breastfeeding (22.7 and 15%, respectively), the incidence of breastfeeding is much lower for black women, 23.0 and 7.0% immediately and 5–6 months after delivery, respectively [10]. Furthermore, there are distinct demographic differences with regard to the extent of breastfeeding in the USA, the more educated and affluent regions of the West Coast and Mountain areas have a much higher breastfeeding incidence than the less affluent East South Central regions.

From these surveys one may conclude that, in general, the population who breastfeeds in the USA is more affluent, better educated and, thus, well nourished.

Milk Composition

Table 1 presents estimates of the concentration of nutrients in mature human milk (i.e. after the colostrum stage of early lactation). The nutrients in milk have three different sources: some are synthesized de novo within the lactating mammary gland (e.g. medium chain fatty acids), some are modified within the mammary secretory cells from precursors taken up from the circulation (e.g. long chain fatty acids incorporated into milk triglycerides) and the third category are transferred into milk directly from plasma (e.g. vitamins and minerals, which are taken up from the circulation, without modification within the mammary gland). Endocrine changes specific to lactation are known to modulate mammary gland function and milk composition. For example, the marked increase in mammary gland lipoprotein lipase activity and its concomitant sharp decrease in adipose tissue at the onset of lactation [11] both under the effect of prolactin [12], result in channeling of dietary fat to the mammary gland for incorporation into milk lipids rather than deposition in adipose tissue.

On the other hand, lactose is synthesized within the mammary gland from the precursors galactose and glucose provided from the circulation. Lactose, present at a concentration of 7 g/dl is the most constant nutrient of human milk and is not affected by maternal nutrition. Proteins amount to about 0.9 g/dl in mature milk [13] and, based on studies in industrialized

Table 1. Estimates of the concentrations of nutrients in mature human milk

Nutrients major, minerals		Nutrients vitamins	
Major nutrients	g/l	Fat soluble	µg/l
Lactose	72.0 ± 2.5	vitamin A, RE	670 ± 200 (2,230 IU)
Protein	10.5 ± 2.0	vitamin D	0.55 ± 0.10
Fat	39.0 + 4.0	vitamin E	2,300 + 1,000
Minerals		vitamin K	2.1 ± 0.1
Macronutrient elements	mg/l	Water soluble	µg/l
Calcium	280 ± 26	vitamin B_6	93,000 ± 8,000
Chloride	420 ± 60	vitamin B_{12}	0.97
Magnesium	35 ± 2	biotin	4 ± 1
Phosphorus	140 ± 22	vitamin C	40,000 ± 10,000
Potassium	525 ± 35	folate	85 ± 37
Trace elements	µg/l	niacin	1,500 ± 200
Chromium	50 ± 5	pantothenic acid	1,800 ± 200
Copper	250 ± 30	riboflavin	350 ± 25
Fluoride	16 ± 5	thiamin	210 ± 35
Iodine	110 ± 40		
Iron	300 ± 100		
Manganese	6 ± 2		
Molybdenum	NR		
Selenium	20 ± 5		
Zinc	1,200 ± 200		

Data (means ± SD) are taken from Hamosh et al. [10, p116].
RE = Retinol equivalents; IU = international units; NR = not reported.

and developing countries, seem to be unaffected by maternal diet or body composition [10]. There is, however, limited information that seems to indicate that short-term high protein diets increase the protein and nonprotein nitrogen content of human milk [14], whereas limited maternal food supply leads to lower milk protein levels [14–16].

The majority of milk proteins serve in the immune and nonimmune protection of the newborn from infection. These proteins, immunoglobulins A, G, and M, lactoferrin and lysozyme have various functions in the newborn [17]. At present there is no clear-cut information about the effect of maternal nutritional status on these protective proteins in milk. Some studies suggest the lack of such an effect [18, 19] while others indicate that malnutrition decreases the concentration of these proteins in human milk [20, 21]. Whereas

Table 2. Factors that affect milk fat content and composition

Variable	Change
Gestation	LC-PUFA higher in preterm* and transitional milk
Lactation	phospholipid, cholesterol higher in colostrum (preterm* > term*)
Parity	P10+: lower endogenous synthesis of FA (C6-C16)
Volume	low milk fat concentration associated with high volume
Feed	fat: fore < mid < hind milk
Diet	
High CH	increase in endogenous synthesis of FA (C6-C16)
Low calorie	increase in palmitic acid (C16)
Pregnancy weight gain	positively associated with milk fat content

* Preterm, term refer to milk or colostrum of women who deliver prematurely or at term. FA = Fatty acids; CH = carbohydrate; LC-PUFA = long chain polyunsaturated fatty acids.

the amount and composition of carbohydrate and protein remain relatively constant in mature human milk, the composition of fat is highly variable and to a large extent is immediately (within several hours) affected by maternal nutrition.

Table 2 lists the factors that affect milk fat content and composition. Length of gestation and length of lactation affect especially the content of the lipids that constitute the milk fat globule membrane, phospholipid and cholesterol [22]. The latter are higher in the early stage of lactation (colostrum and transitional milk) because the milk fat globules are much smaller than in mature milk [23, 24] and therefore the total 'membrane' lipid level is higher in milk. Long chain polyunsaturated fatty acids (C20:4ω6 and C22:6ω3) are essential for growth, brain development and retinal function [25]. These fatty acids are stored in the fetus only in the last trimester of pregnancy, therefore preterm infants depend exclusively on human milk for the provision of these essential nutrients. It has recently been shown that the level of ω3 and ω6 LC-PUFA is lower in pregnant and lactating women [26], suggesting preferential transfer of these essential fatty acids from mother to fetus or to the newborn through milk, even at the cost of possible depletion of maternal reserves [26]. It has long been known that milk fat content changes drastically during each feed [27, 28] and that milk fat composition is markedly affected by the maternal diet [29]. More recent studies have shown that the mechanism for endogenous synthesis of fatty acids seems to be exhausted in women of very high parity [30], that infants who receive low fat-containing milk tend to nurse more frequently and for longer time periods thereby causing an increase in milk volume [31], and that there is a strong positive relationship between weight gain during pregnancy and milk fat content [32].

Table 3 lists the relationship between maternal nutrient intake, milk composition and the effect on the infant.

As mentioned above and as stated in table 3, the vitamin content of human milk is dependent upon the mother's vitamin status, and when maternal intake of specific vitamins is chronically low, these vitamins will also be present at lower levels in the milk. With supplementation of vitamins, levels in milk also increase. Water-soluble vitamins in milk are generally more responsive to maternal dietary intake than fat-soluble vitamins [10]. The relationship between maternal intake and milk concentration varies, however, among vitamins. For example, excess vitamin C does not further increase the level in milk (above that associated with adequate intake), whereas milk vitamin B_6 levels continue to rise with higher intakes; milk folate levels remain normal even at the expense of maternal folate stores and do not decrease until the latter are depleted [10]. Based on the concentrations of fat-soluble vitamins in human milk and infant needs it is advised that in the USA all newborns should receive 0.5–1.0 mg vitamin K by injection or 1.0–2.0 mg orally immediately after birth [10]; infants should receive 5.0–7.5 μg vitamin D per day if exposure to sunlight seems inadequate.

The concentration of trace minerals varies as a function of length of lactation (iron, copper, zinc, selenium). While concentration in milk seems to be independent of maternal nutrition for iron [33, 34], or fluoride [35], that of maganese [36], iodine [37] and selenium [38] is. Iodine is unique among trace elements in that it is avidly accumulated by the mammary gland.

It is recommended that breastfed infants be supplemented with fluoride if the water supply for their family is low (<0.3 ppm). Because of the high bioavailability of iron in human milk, exclusively breastfed infants do not need iron supplements during the first 6 months of life. However, when supplementary foods are introduced (as recommended after 4–6 months of exclusive breastfeeding) iron should be supplemented to the infant's nutrition.

Concerns about Specific Nutrients and Specific Populations with Regard to Appropriate Nutrition during Lactation

In the United States there are concerns that the following nutrients might not meet the RDA for lactating women:

Calcium: it is difficult to consume 1,200 mg daily especially if dairy products are not an important part of the diet.

Table 3. Possible influences of maternal intake on the nutrient composition of human milk and nutrients for which clinical deficiency is recognizable in infants

Nutrient or nutrient class	Effect of maternal intake on milk composition	Recognizable nutritional deficiency in breastfed infants
Macronutrients		
Proteins	+	unknown[1]
Lipids	+[2]	unknown
Lactose	0	unknown
Minerals		
Calcium	0	unknown
Phosphorus	0	unknown
Magnesium	0	unknown
Sodium	0	unknown
Potassium	0	unknown
Chlorine	0	unknown
Iron	0	yes[3]
Copper	0	unknown
Zinc	+, 0	unknown
Manganese	+	unknown
Selenium	+	unknown
Iodine	+	yes
Fluoride	+	unknown
Vitamins		
Vitamin C	+	yes
Thiamin	+	yes
Riboflavin	+	unknown
Niacin	+	unknown
Pantothenic acid	+	unknown
Vitamin B_6	+	yes
Biotin	+	yes
Folate	+	yes
Vitamin B_{12}	+	yes
Vitamin A	+	yes
Vitamin D	+	yes
Vitamin E	+	yes
Vitamin K	+	yes[4]

Data from Hamosh et al. [10], with permission.

+ = A positive effect of intake on nutrient content of the milk; the magnitude of the effect varies widely among nutrients; 0 = no known effect of intake on nutrient content of the milk.

[1] Evidence is not sufficiently conclusive to categorize as 'no'.

[2] Effect appears to be on type of fatty acids present but not on total content of triglycerides or cholesterol in the milk.

[3] Deficiency is not related to maternal intake.

[4] Maternal intake is not the primary determinant of the infant's vitamin K status.

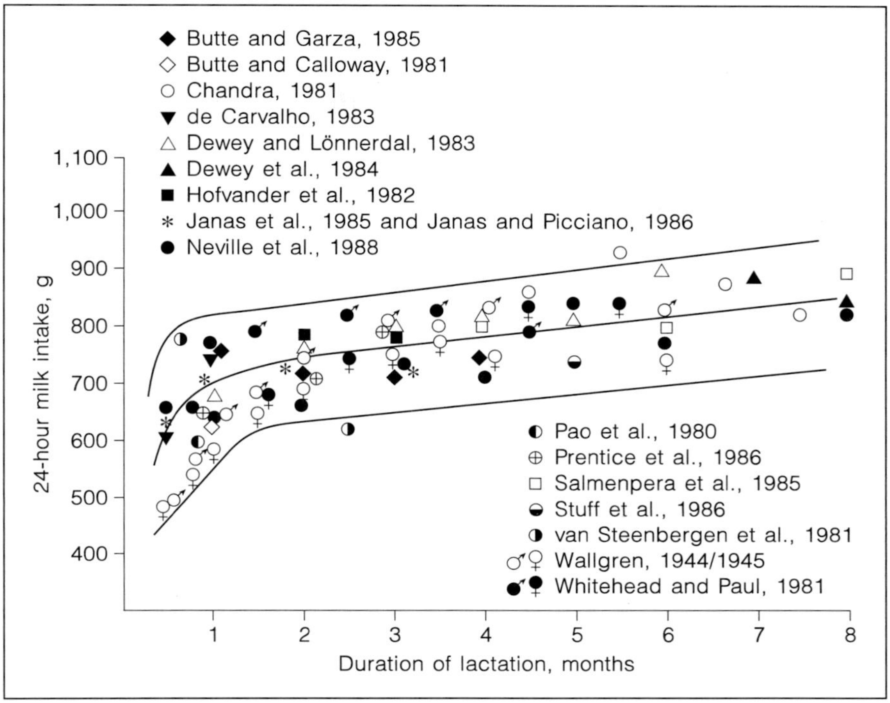

Fig. 1. Milk intake during established lactation. For references to individual studies see Hamosh et al. [10] and Neville et al. [42]. From Hamosh et al. [10], with permission.

Zinc: intake seems to be low, although plasma zinc concentrations are similar in lactating and nonlactating women in the USA [38].

Magnesium: intake in lactating women is marginal.

Vitamin B_6, thiamine and folate are of concern in the nutrition of lactating women. Populations at risk for inadequate intake of certain nutrients during lactation are adolescents (less iron intake), poor adolescents (low vitamin A intake), black women (lower calcium, magnesium and vitamin A intake than the general population of nursing women), women with low income (less calcium and vitamin A consumption), complete vegetarians (low vitamin B_{12} intake).

Among undesired practices that might affect milk production it is recommended that lactating women refrain from smoking or abuse of illegal drugs and that alcohol and coffee be consumed only in moderation (i.e. alcohol not more than 0.5 g/kg/day and coffee no more than 2 cups/day). Indeed, a recent study shows that ethanol consumption (0.3 g/kg body weight) by nursing

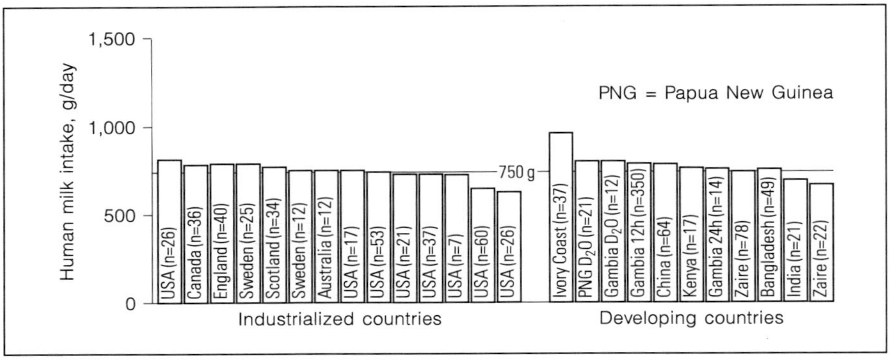

Fig. 2. Milk intake at 3 months of age in industrialized and developing countries [10, 43]. From Hamosh et al. [10], with permission.

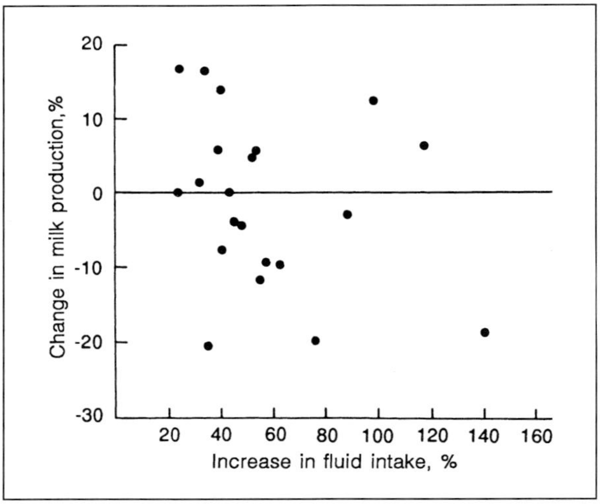

Fig. 3. Relationship between increase in maternal fluid intake for 3 days and percent change in milk production [10, 44]. From Hamosh et al. [10], with permission.

mothers has an immediate effect on the odor of the milk and on the feeding behavior of the infants, who consumed significantly less milk within 3 h after their mothers drank alcohol [40].

Milk consumption can, however, be positively affected by certain ingredients in the mother's food, such as garlic, which has recently been shown to affect milk odor and to increase milk consumption by infants [41].

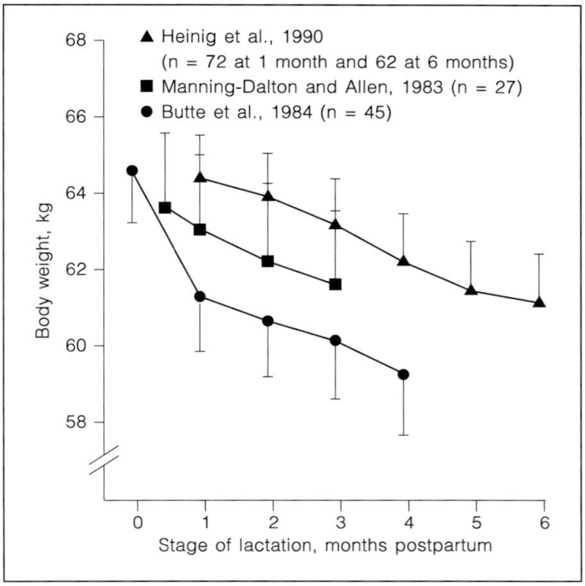

Fig. 4. Change in maternal body weight (mean ± SD) during lactation in American women. From Hamosh et al. [10], with permission.

Milk Volume and Maternal Nutrition

Milk volume seems to be relatively constant irrespective of maternal nutritional status. Figure 1 shows the general pattern of the daily volume of milk intake of infants from birth to the age of 8 months. The data that summarize 16 studies in several industrialized countries show that, in general, healthy infants have an average daily intake of 750–800 ml milk within the first 4–5 months after birth [10, 42]. There is, however, a much wider range of intake volumes from 450 to 1,200 ml/day [10, 43]. Similar data are reported from developing countries [10, 43] (fig. 2), although the maternal nutritional status might be subject to seasonal variation and might, in general, be less adequate than in industrialized countries. Increasing the intake of fluid does not seem to affect milk volume [42] (fig. 3); therefore, lactating women should maintain adequate fluid intake but should be aware that 'fluids consumed in excess of natural thirst have no effect on milk volume' [10].

Women should be encouraged to consume at least 1,800 kcal/day and should be discouraged from dieting with the aim of excessive weight loss during lactation, although a recent study shows that careful dieting (i.e. reduction of

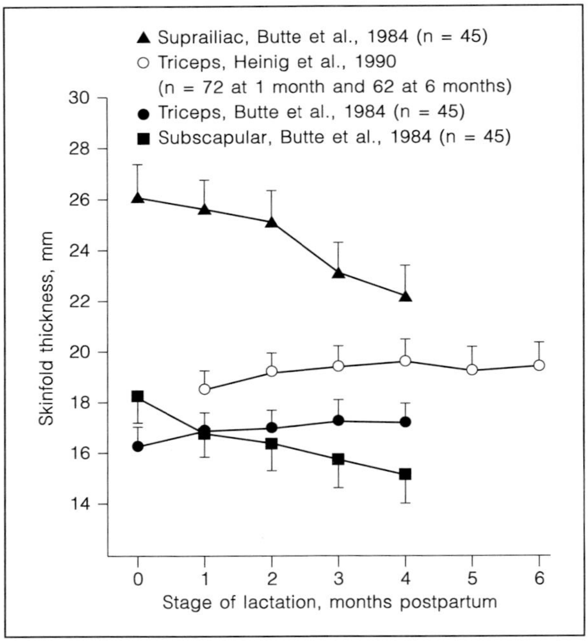

Fig. 5. Change in skinfold thickness (mean ± SD) during lactation. From Hamosh et al. [10], with permission.

energy intake and adequate protein, vitamin, calcium and iron intake) that resulted in a loss of 4.8 kg during a 10-week period, did not affect milk volume and composition nor infant growth [45].

Aerobic exercise (at a level 60–70% of the heart rate reserve) 5 days weekly for a period of 12 weeks did not adversely affect milk volume and composition (fat, carbohydrate or protein content) or infant milk consumption and growth [46]. However, while exercise in well-nourished women (who compensate for the higher energy expenditure with higher energy intake [46]) does not seem to affect lactation performance, strenuous exercise in marginally nourished women might have different effects on lactation performance.

There are currently no established standards for assessing nutritional status of lactating women [10]. On average, lactating women lose 0.6–0.8 kg/month during the first 4–6 months postpartum (fig. 4); there is, however, wide variation among women and some may gain weight during lactation [10]. Skin fold thicknesses (subscapular and suprailiac but not triceps) decrease during the first 4–6 months of lactation [10] (fig. 5). However, while women in affluent

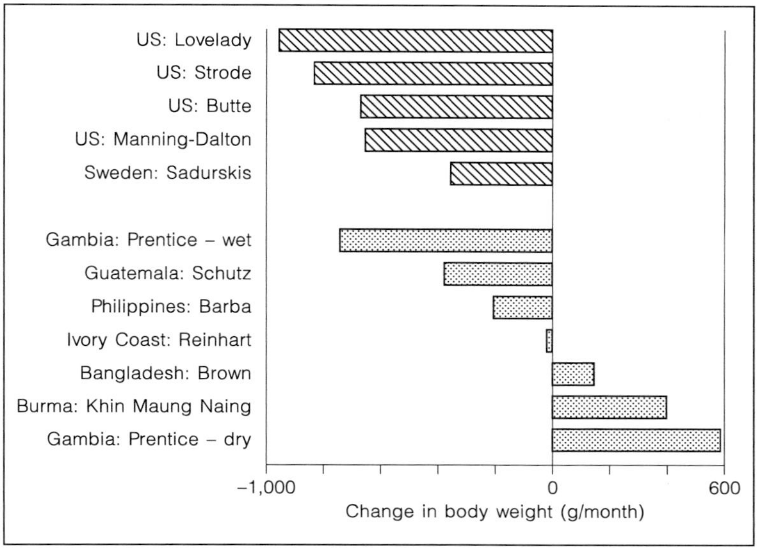

Fig. 6. Change in body weight during lactation of women in industrialized and developing countries. From Brown and Dewey [47], with permission.

societies tend to lose weight during lactation, those in developing countries maintain constant weight or gain weight [47] (fig. 6). The latter consume less energy than the former, but because of smaller size actually consume more energy per kilogram body weight [47]. The authors suggest that the difference between these two groups of women could be attributed to lower weight gain during pregnancy rendering women in developing countries more dependent upon postpartum weight gain for the maintenance of lactation [47]. Although 'nutrition is robust' and is adequate even in marginally nourished women, efforts should be made that all lactating women be adequately nourished in order to prevent depletion of maternal stores. A recently published guide provides practical information on means to assess and counsel pregnant and lactating women on proper nutrition during these periods of their life [48].

Acknowledgement

The excellent help of Anne M. Friedrich in the preparation of the manuscript is greatly acknowledged.

References

1 Martinez GA, Krieger FW: 1984 Milk-feeding patterns in the United States. Pediatrics 1985;76: 1004–1008.

2 FASEB (Federation of American Societies for Experimental Biology): Proceedings of the Federation of American Societies for Experimental Biology. J Pediatr Gastroenterol Nutr 1984;3.

3 Jensen RG, Neville MC (eds): Human Lactation: Milk Components and Methodologies. New York, Plenum Press, 1985, 307 pp.

4 Hamosh M, Goldman AS (eds): Human Lactation. 2. Maternal and Environmental Factors, New York, Plenum Press, 1986, 657 pp.

5 Goldman AS, Atkinson SA, Hanson LA: Human Lactation. 3. The Effects of Human Milk on the Recipient Infant. New York, Plenum Press, 1987, 400 pp.

6 Picciano MF, Lonnerdal B: Mechanisms Regulating Lactation and Infant Nutrient Utilization. New York, Wiley-Liss, 1992, 463 pp.

7 Mestecky J, Blair C, Ogra PL: Immunology of Milk and the Neonate. New York, Plenum Press, 1991, 483 pp.

8 Jensen RG, Thompson MP (eds): Handbook of Milk Composition. New York, Academic Press, 1995.

9 King JC, Allen L, Butte NF, Chez RA, Haas JD, Kleinman JC, Kramer MS, Lederman SA, Berkowitz GS, Dallman P, Hambidge KM, Kirskey A, Niebyl JR, Sparks J: Nutrition during Lactation. Washington, National Academy Press, 1990, 468 pp.

10 Hamosh M, Dewey KG, Garza C, Goldman AS, Lawrence RA, Picciano MF, Quandt SA, Rasmussen KM, Rush D: Nutrition during Lactation. Washington, National Academy Press, 1990, 309 pp.

11 Hamosh M, Clary TR, Chernick SS, Scow RO: Lipoprotein lipase activity of adipose and mammary tissue and plasma triglyceride in pregnant and lactating rats. Biochim Biophys Acta 1970;210: 473–482.

12 Zinder O, Hamosh M, Fleck TRC, Scow RO: Effect of prolactin on lipoprotein lipase in mammary gland and adipose tissue of rats. Am J Physiol 1974;226:744–748.

13 Lönnerdal B, Forsum E, Hambraeus L: A longitudinal study of the protein, nitrogen, and lactose contents of human milk from Swedish well-nourished mothers. Am J Clin Nutr 1976;29:1127–1133.

14 Deb AK, Cama HR: Studies on human lactation. Dietary nitrogen utilization during lactation, and distribution of nitrogen in mother's milk. Br J Nutr 1962;16:65–73.

15 Lindblad BS, Rahimtoola RJ: A pilot study of the quality of human milk in a lower socioeconomic group in Karachi, Pakistan. Acta Paediatr Scand 1974;63:125–128.

16 Wurtman JJ, Fernstrom JD: Free amino acid, protein, and fat contents of breast milk from Guatemalan mothers consuming a corn-based diet. Early Hum Dev 1979;3:67–77.

17 Hamosh M: Human milk composition and function in the infant. Semin Pediatr Gastroenterol Nutr 1992;3:4–8.

18 Cruz JR, Carlsson B, Garcia B: Studies in human milk. III. Secretory IgA quantity and antibody levels against *Escherichia coli* in colostrum and milk from under privileged and privileged mothers. Pediatr Res 1982;16:272–276.

19 Reddy V, Srikantia SG: Interaction of nutrition and the immune response. Indian J Med Res 1978; 66:48–57.

20 Miranda R, Saravia NG, Ackerman R, Murphy N, Berman S, McMurray DN: Effect of maternal nutritional status on immunological substances in human colostrum and milk. Am J Clin Nutr 1983;37:632–640.

21 Roberston DM, Carlsson B, Coffman K, Han-Zoric M, Salil F, Jones C, Hanson LA: Avidity of IgA antibody to *Escherichia coli* polysaccharide and diphtheria toxin in breast milk from Swedish and Pakistani mothers. Scand J Immunol 1988;28:783–789.

22 Bitman J, Wood DL, Hamosh M, Hamosh P, Mehta NR: Comparison of the lipid composition of breast milk from mothers of term and preterm infants. Am J Clin Nutr 1983;38:300–312.

23 Ruegg M, Blanc B: The fat globule size distribution in human milk. Biochim Biophys Acta 1981; 666:7–14.

24 Simonin C, Ruegg M, Sidiropoulos D: Comparison of the fat content and fat globule size distribution of breast milk from mothers delivering term and preterm. Am J Clin Nutr 1984;40:820–826.
25 Innis SM: Essential fatty acids in growth and development. Prog Lipid Res 1991;30:39–103.
26 Holman RT, Johnson SB, Ogburn PL: Deficiency of essential fatty acids and membrane fluidity during pregnancy and lactation. Proc Natl Acad Sci USA 1991;88:4835–4839.
27 Hytten FE: Clinical and chemical studies in human lactation: Variations in major constituents during a feeding. Br Med J 1954;i:176–179.
28 Macy IG, Nims B, Broen M, Hunscher HA: Human milk studies. VII. Chemical analysis of milk representative of the entire first and last halves of the nursing period. Am J Dis Child 1931;42: 569–589.
29 Insull W Jr, Hirsch J, James T, Ahrens EH Jr: The fatty acids of human milk. II. Alterations produced by manipulation of caloric balance and exchange of dietary fats. J Clin Invest 1959;38: 443–450.
30 Prentice A, Jarjou LM, Drury PJ, Dewit O, Crawford MA: Breast-milk fatty acids of rural Gambian mothers: Effects of diet and maternal parity. J Pediatr Gastroenterol Nutr 1989;8:486–490.
31 Tyson J, Burchfield J, Sentance F, Mize C, Uauy R, Eastburn J: Adaptation of feeding to a low fat yield in breast milk. Pediatrics 1992;89:215–220.
32 Michaelsen KF, Larsen PS, Thomsen BL, Samuelson G: The Copenhagen cohort study on infant nutrition and growth: Breastmilk intake, human milk macronutrient content, and influencing factors. Am J Clin Nutr 1994;59:600–611.
33 Dallman PR: Iron deficiency in the weanling: A nutritional problem on the way to resolution. Acta Paediatr Scand 1986;323(suppl):59–67.
34 Siimes MA, Salmenperä L, Perheentupa J: Exclusive breast-feeding for 9 months: Risk of iron deficiency. J Pediatr 1984;104:196–199.
35 Ekstrand J, Spak CJ, Falch J, Afseth J, Ulvestad H: Distribution of fluoride to human breast milk: Following intake of high doses of fluoride. Caries Res 1984;18:93–95.
36 Vuori E, Mäkinen SM, Kara R, Kuitunen P: The effects of the dietary intakes of copper, iron, manganese, and zinc on the trace element content of human milk. Am J Clin Nutr 1980;33:227–231.
37 Gushurst CA, Mueller JA, Green JA, Sedor F: Breast milk iodide: Reassessment in the 1980s. Pediatrics 1984;73:354–357.
38 Mannan S, Picciano MF: Influence of maternal selenium status on human milk selenium concentration and glutathione peroxidase activity. Am J Clin Nutr 1987;46:95–100.
39 Moser PB, Reynolds RD: Dietary zinc intake and zinc concentrations of plasma, erythrocytes, and breast milk in antepartum and postpartum lactating and nonlactating women: A longitudinal study. Am J Clin Nutr 1983;38:101–108.
40 Mennella JA, Beauchamp GK: The transfer of alcohol to human milk: Effect of flavor on the infant's behavior. N Engl J Med 1991;325:981–985.
41 Mennella JA, Beauchamp GK: Maternal diet alters the sensory qualities of human milk and behavior of the nursling. Pediatrics 1991;88:737–747.
42 Neville MC, Keller R, Seacat J, Lutes V, Neifert M, Casey C, Allen J, Archer P: Studies in human lactation: Milk volumes in lactating women during the onset of lactation and full lactation. Am J Clin Nutr 1988;48:1375–1386.
43 Prentice A, Paul A, Prentice A, Black A, Cole T, Whitehead R: Crosscultural differences in lactational performance; in Hamosh M, Goldman AS (eds): Human Lactation. 2. Maternal and Environmental Factors. New York, Plenum Press, 1986, pp 13–44.
44 Dusdieker LB, Booth BM, Stumbo PJ, Eichenberger JM: Effect of supplemental fluids on human milk production. J Pediatr 1985;106:207–211.
45 Dusdieker LB, Hemingway DL, Stumbo PJ: Is milk production impaired by dieting during lactation? Am J Clin Nutr 1994;59:833–840.
46 Dewey KG, Lovelady CL, Nommsen-Rivers LA, McCrory MA, Lönnerdal B: A randomized study of the effects of aerobic exercise by lactating women on breast-milk volume and composition. N Engl J Med 1994;330:449–453.

47 Brown KH, Dewey KG: Relationship between maternal nutritional status and milk energy output of women in developing countries; in Picciano MF, Lönnerdal B (eds): Mechanisms Regulating Lactation and Infant Nutrient Utilization. New York, Wiley-Liss, 1992, pp 77–95.
48 Olson C, Wilson J, Abrams B, Alton I, Chez RA, Dallman P, Mullen PD, Paine LL, Smith MA, Windsor RA: Nutrition during Pregnancy and Lactation, an Implementation Guide. Washington, Academy Press, 1992, pp 133.

Prof. Margit Hamosh, Chief, Division of Developmental Biology and Nutrition,
Department of Pediatrics, Georgetown University Medical Center,
3800 Reservoir Rd. NW, Washington D.C. (USA)

Porrini M, Walter P (eds): Nutrition in Pregnancy and Growth.
Bibl Nutr Dieta. Basel, Karger, 1996, No 53, pp 37–44

Composition of Human Milk: Nutritional Aspects

Leif Hambræus

Department of Nutrition, Faculty of Medicine, University of Uppsala, Sweden

Introduction

The new concept of functional foods, which was recently introduced by Japanese scientists has stimulated the discussion in nutrition and food science regarding the physiological role of certain components in food. Although it may look like a new perspective on nutritional matters, it could also be said that the concept of functional foods by no means represents anything new. Thus, mammalian milk must by all means represent *the* functional food as it contains a number of macro- and micronutrients which have specific functional roles for the development and growth of the offspring. Further indications of the metabolic and nutritional implication of the composition of milk are obtained from the fact that there are great variations in the nutrient content in milk from various species [1].

Why Should Milk Have a Specific Nutritional Message?

One of the major characteristics of the nutritional situation during the neonatal period in mammals is the fact that the offspring has to rely upon one single source of nutrients – milk, which means that there is no possibility to compensate for any lack of nutrients in the diet. At the same time there is a very pronounced growth and maturation of the tissues during the neonatal period which creates an increased and specific demand for essential nutrients. A reduced tolerance for deviations in food intake due to the immaturity of the liver and kidneys, both of which play essential roles in the metabolism and excretion of metabolites, however, complicates the nutritional situation.

Consequently, it has rightly or wrongly been assumed that the composition of the milk of any mammalian species is the definite indicator of the optimum composition of nutrients required during the neonatal period of that species.

Grouping of Nutrients in Milk

The nutritional aspects of milk components can essentially be divided into two sections: a macronutrient aspect and a micronutrient or specific physiological aspect. The macronutrient aspects comprise essentially the role of major dietary components used more or less as nonspecific nutrient sources, i.e. energy and protein, which may play essential roles in anabolic processes as a source of material for growth and repair of tissues during convalescence. However, there is a growing interest nowadays devoted to the specific physiological aspects of the micronutrients, i.e. trace elements, vitamins, hormones and growth factors and their roles for metabolic, endocrine and immunological functions in the body [2].

The Protein Dilemma: Nutritional versus Physiological Roles of Milk Proteins

The milk proteins and their potential physiological function [3] are of great interest. It could be said that the milk proteins have three roles to play: (1) A conventional dietary source of essential amino acids and nitrogen for protein synthesis; (2) a source of protein for specific physiological functions, i.e. defence against microbial infections; (3) growth factors and modulators (essentially low-molecular proteins occurring in minor amounts).

The macronutrient aspects of milk protein must take into account the potential and specific physiological role of certain protein fractions. If some of the proteins are considered to exert their specific physiological roles they cannot be hydrolyzed in the gastrointestinal tract as well. This statement has at least two dimensions:

(1) A protein determination based only on total nitrogen analysis might be greatly misleading for calculation of the nutritionally available milk proteins as such analytical data does not differ between protein nitrogen and nonprotein nitrogen (NPN). In the case of human milk 25% of the nitrogen is derived from NPN instead of 5% as is the case in bovine milk [2]. Without this correction for NPN, human milk has been assumed to contain 1.1–1.2 g protein/100 ml, which unfortunately is still often given in the textbooks, instead of 0.8–0.9 which is the real value, i.e. an overestimation with 25% [4]. Another

most interesting question in this respect is of course the nutritional value of the NPN in human milk [5]. This is still not fully evaluated and there are conflicting data in the literature and more critical studies are needed.

(2) A determination of true protein content leading to a value of 0.8–0.9 g/100 ml may still represent an overestimation if it does not take into account whether these proteins are nutritionally available or not. This is especially valid with respect to human milk where proteins such as secretory IgA, lactoferrin and lysozyme all constitute a considerable amount of the whey protein fraction, which is the dominating one in human milk. Since human milk contains substantial amounts of physiologically active proteins which may exert their physiological effects only if they are kept intact throughout the gastrointestinal tract, they cannot be considered as available dietary sources of amino acids as well. In practice, human milk may consequently have a content of nutritionally available protein as low as 0.5–0.6g/100 ml constituting only 4% or less of the total energy content [4, 6]. The meaning of the extremely low protein energy % in human milk is still not known. It is still a matter of discussion which protein level is recommended on the basis of the protein content in human milk [7, 8].

Referring to the reasoning above, all data regarding protein requirement in man during the neonatal period may be overestimates with about 100%. Even in the latest version of the FAO/WHO/UNU [9] on energy and protein requirement, they are based on the assumption that human milk contains 1.1–1.2g/100 ml instead of 0.5–0.6. They furthermore state:

'It is recognized that human milk contains about 20% of total nitrogen as non-amino nitrogen but following the usual convention we have assumed that this nitrogen fraction is utilized.'

A too high protein intake may be potentially harmful giving rise not only to kidney damage [10] and diabetes [11, 12] but also the negative effect of sulphur amino acids, especially homocysteine on the cardiovascular system is today much more discussed [13].

Milk as Energy Source: Which Fat Energy Percent is Recommendable?

From the energy point of view (table 1), the nutritionally interesting aspect would essentially be to evaluate what the energy percentage distribution between the energy-yielding nutrients in human milk might indicate for future life. Interestingly, human milk is quite extreme versus not only cow's milk but also the characteristics of the diet in later life. It is still not very clear what such a difference in the distribution of energy between the various energy-yielding macronutrients will mean in practice for the future outcome and

Table 1. Energy distribution in human and bovine milk versus adult western diet

Energy source	Human milk	Cow's milk	Western diet
Protein	6	21	12
Fat	52	50	38
Carbohydrate	42	29	50

metabolic programming of the offspring. How fast should we change dietary fat intake from 50 E% in human milk to <30 E% in the dietary recommendations of adults? How should we interpret the effect of maternal diet on milk fat composition for the outcome of the future health of the offspring?

Effect of Maternal Diet and Nutritional Status on Milk Composition

One of the most important environmental factors must be the effect of the nutritional status of the mother and her diet on milk composition. This might also indirectly give us some information about the necessity of a homeostatic mechanism to keep the nutritional quality within certain limits. But the question is then to what extent should this be taken as a cause to manipulate the mother's diet? The concentrations of fatty acids as well as the content of vitamins, fat-soluble as well as water-soluble, are significantly affected by the maternal diet. On the other hand lactose content as well as the content of minerals and trace elements seem to be relatively resistant to varying intakes of the mother [14].

The impact of maternal dietary protein intake on milk composition has been difficult to evaluate due to the methodological problems with protein determination as mentioned earlier. In general, it seems that malnutrition has little if any effect on protein concentration in human milk. Likewise protein supplementation of malnourished women does not appear to have any pronounced effect on milk protein concentration. However, a better nutritional status of the mother may have an impact on the total volume of milk produced.

The Fat Dilemma: Which Fat Quality Is Recommendable?

Lipids constitute about 3–5% of milk and 98% are triacylglycerols, thereby constituting 50–60% of the energy. Only 8–10 fatty acids constitute more than 1% each. However, minor lipid compounds such as phospholipids, sterols and fat-soluble vitamins are of course also of physiological importance although they occur in minor amounts. Thus, this situation resembles the balance between nutritional value of protein and physiological functions of specific protein fractions.

Not only the total fat content but also the fatty acid composition of human milk fat is affected by the maternal diet and this has been the focus in a number of studies. Human milk contains a substantial amount of unsaturated fatty acids, i.e. oleic acid and linoleic acid, while the concentration of saturated fatty acids is low. A relatively high proportion represents short-chain (C4-C6) or medium-chain (C8-C12) fatty acids. Monounsaturated constitute about 30% while polyunsaturated are less than 3%. Interestingly, the fat content seems to be influenced by the mother's diet and a marked elevation in the MCT fatty acids is reported to occur on a low-fat diet. The linoleic acid content is furthermore reported to show the most pronounced variation with maternal diet being extremely high in vegetarians and vegans.

Many investigators have attempted to correlate poor maternal nutritional status with milk fat concentration but there seems to be less impact than assumed. This might be due to the increased use of depot fat to synthesize milk fat during energy deficit. The effect on fatty acid pattern is not only influenced by the fatty acid composition of the maternal diet in itself, but also to the balance between energy-yielding nutrients. Thus a low-fat, high-energy diet leading to synthesis of fatty acids from carbohydrate leads to higher content of saturated fatty acids but when low-fat, low-energy diet was fed, the fatty acid pattern resembled that of depot fat. Several studies have shown that the differences in fatty acid composition of milk fat in various population groups may be due to their dietary habits. Should this be considered as an indication of a trial of adaptation to the future diet of the offspring? Or should it be indication of the necessity to change dietary habits of the mother for the future outcome of health of the infant?

Lactoferrin – A Functional Protein of Specific Interest

Since lactoferrin is a good illustration of the various concepts regarding the nutritional and physiological roles of certain compounds in the milk this will be used as an example. It is one of the major protein components in human whey, in the human colostral whey even the dominating one as this contains as much as 6 mg/ml. Lactoferrin is produced also in other fluids produced in exocrine glands, such as tears, saliva and pancreatic juice. It is also a component of the neutrophiles secretory granules.

Lactoferrin has multiple biological functions [15], not only in the defense system against gastrointestinal infections. It also has other biological functions, i.e. as growth factor, iron transporter and in the defence against free radical formation, all of which seem to be of clinical significance. The growth and maturation of the gastrointestinal tract is more rapid in animals and humans fed lactoferrin-containing milk than those on commercial products. This stimulatory effect is not related to the iron saturation. The newly described bactericidal effect of a peptide from lactoferrin, lactoferricin, is of special interest. Lactoferrin also occurs in neutrophils during inflammation and is activated during inflammation. This fits with the theory that lactoferrin plays a role in reducing the free radical effect during the inflammatory process by transferring iron to the RES to be incorporated in ferritin. Another characteristic of lactoferrin which may have a significant role of clinical implication is that it can bind to a variety of normal and leukemic cells and that lactoferrin released by neutrophils plays a role in modulating the immune and inflammatory responses. This may be secondary to aggregation of neutrophils and their adhesion to epithelial cells which may be a factor that causes neutrophils to accumulate at inflammatory sites. The iron-saturated form of lactoferrin is a highly active form of the active inhibitor of myelopoesis which may indicate that lactoferrin could be of use in the treatment of leukemia.

In milk, as in other secretions, lactoferrin appears mostly in the apolactoferrin form, i.e. iron-free as its saturation is only 8–10%. It is in this iron-free form, apolactoferrin, that it exerts its bacteriostatic properties as it binds adventitious iron so strongly that it deprives the bacteria for the iron needed for their growth. Interestingly, despite its high affinity to iron, only 30% of the iron in human milk is bound to lactoferrin. One much discussed question has been the absorption of lactoferrin from mother's milk in the intestines of the nursing infant. Firstly, there are conflicting results regarding the absorption of lactoferrin as such, and, secondly, it has been shown to be excreted in the infants' urine. So far, however, there is no published evidence of an in vivo activity of lactoferrin.

Another matter of concern is the role of lactoferrin for the bioavailability of iron in milk. The extremely high bioavailability reported of iron in human milk [16] has been assumed to be due to the role of lactoferrin. Interesting recent studies illustrate the role of lactoferrin as a transporter and that there should be special lactoferrin receptors in the gut mucosa [15]. Considerable interest has been devoted to the occurrence of specific lactoferrin receptors. The exact role of the receptors is still not known but it may have an impact on the uptake of iron. However, it seems as these receptors are very species specific and consequently the potential role of lactoferrin seems to be very species specific. Consequently, bovine lactoferrin might not have the same beneficial effects as those characterising human lactoferrin. Once more we are facing a practical problem of relevance when comparing various milks.

The Specific Use of Lactoferrin in Functional Foods

Since lactoferrin also occurs in bovine milk, although in a much lower concentration, modern food technology has made it possible to isolate and concentrate lactoferrin making it available for pharmacological and other purposes. Of specific interest is the fact that there seems to be a species specificity of lactoferrin which might indicate that also the biological function is species specific. Despite the fact that lactoferrin shows species specificity, there is an international trend toward the addition of bovine lactoferrin to infant formula. It is already mentioned as an example of a component in functional foods in the Japanese report on functional foods. It has already been the case in Japan and there is an interest to do the same in Europe and the USA. This is probably partly due to the fact that lactoferrin today can be produced and purified in significant amounts by ion-exchange chromatography, ultrafiltration and affinity chromatography. Interestingly, the human lactoferrin gene has now been cloned which makes large-scale lactoferrin production possible. Thus, European formula manufacturers have produced a transgenic bull which carries the gene for human lactoferrin. This animal is being used to sire daughters and granddaughters who are expected to produce human milk containing human lactoferrin. The plan is of course to use this milk to manufacture infant formula. Thus, lactoferrin is a good example of what could be a component in functional foods. At the same time, it illustrates that the food industry has taken into account a possible mechanism in their marketing of lactoferrin before it has been objectively shown by a scientifically sound backup.

Conclusion

The nutritional significance of the specific components of human milk is still a matter of great interest. It can be assumed that further knowledge of the macro- as well as microcomponents of breastmilk will give us valuable information regarding the potential use of various milk compounds in pharmaceutical preparations as well as in functional foods.

References

1 Widdowson EM: Lactation and feeding patterns in different species; in Freed DLJ (ed): Health Hazards of Milk: 85–90. Eastbourne, Baillière Tindall/Saunders 1984.

2 Hambræus L: Human milk: Nutritional aspects; in Brunser O, Carrazza F, Gracey M, Nichols B, Senterre J, (eds): Clinical Nutrition in the Young Child. Vevey, Nestec, New York, Raven Press, 1990, pp 289–301.

3 Lönnerdal, B: Biochemistry and physiological function of human milk proteins. Am J Clin Nutr 1985;42:1299–1317.
4 Hambræus L, Fransson GB, Lönnerdal B: Nutritional availability of breast milk protein. Lancet 1984; ii: 167–168.
5 Lönnerdal B: Nutritional importance of non-protein nitrogen; in Räihä NE (ed): Protein Metabolism during Infancy. Nestle Nutrition Workshop Ser. New York, Raven Press, 1994, vol 13, pp 105–116.
6 Van Woelderen B: Changing insight into milk proteins: Some implications. Nutr Abstr Rev 1987; 57:129–134.
7 Lönnerdal B, Zetterström R: Protein content of infant formula: How much and from what age? Acta Paediatr Scand 1988;77:321–325.
8 Räihä NCR: Protein content of human milk from colostrum to mature milk; in Räihä NE (ed): Protein Metabolism during Infancy. Nestle Nutrition Workshop Ser. New York. Raven Press, 1994, pp 87–99.
9 FAO/WHO/UNU Expert consultation: Energy and protein requirements. WHO Techn Rep Ser No. 724, Geneva, WHO, 1985.
10 Brener R, Meyer TW, Hostetter TH: Dietary protein intake and the progressive nature of kidney disease: The role of hemodynamically mediated glomerular injury in the pathogenesis of progressive glomerulosclerosis in aging, renal ablation and intrinsic renal disease. N Engl J Med 1982;307: 652–659.
11 Pedersen MM, Winther E, Mongensen CR: Reducing protein in the diabetic diet. Diabèt Métabol 1990; 16:454–459.
12 British Diabetic Association, Nutrition Subcommittee: Dietary recommendation for people with diabetes: An update for the 1990s. Diabet Med 1992; 9:189–202.
13 Kang S-S, Wong PWK, Malinow MR: Hyperhomocyst(e)inemia as a risk factor for occlusive vascular disease. Ann Rev Nutr 1992;12:279–298.
14 Lönnerdal B: Effects of maternal dietary intake on human milk composition. J Nutr 1986;116: 499–513.
15 Hambræus L, Lönnerdal B: The physiological role of lactoferrin. In Proc. IDF Seminar: Indigenous antimicrobial agents of milk. Int. Dairy Fed., Brussels, Belgium 1994, pp 97–107.
16 Saarinen UM, Siimes MA, Dallman PR: Iron absorption in infants: High bioavailability from human milk, simulated human milk and proprietary formulas. Pediatrics 1977;60:896–900.

Leif Hambræus, MD, Department of Nutrition,
Dag Hammarskjölds väg, 21, S–752 37 Uppsala (Sweden)

Porrini M, Walter P (eds): Nutrition in Pregnancy and Growth.
Bibl Nutr Dieta. Basel, Karger, 1996, No 53, pp 45–54

Linear Growth in Children

J. C. Waterlow

Department of Public Health & Policy, London School of Hygiene and Tropical Medicine, London, UK

Introduction

The Cambridge physiologist Sir Joseph Barcroft used to tell us that if you want to understand a process or an equilibrium you must disturb it; so the disturbances imposed by nature may help us to better understand the factors that control normal linear growth. I therefore make no apology to this assembly of European nutritionists for concentrating in this talk on growth in third-world children. The reduction in linear growth that they so commonly show is undoubtedly an effect of a deprived environment, but beyond that we know virtually nothing about the cause, in biological as opposed to social terms.

Some 20 years ago, when I proposed a system of classifying malnourished children, I called these children with a deficit in linear growth 'stunted' [1]. It has been suggested that this term is offensive to people in the third world, but I have not found it to be so. Moreover, we can usefully distinguish between 'stunted' as a state and 'stunting' as a process. Some people prefer to use the word 'short', but to me 'short' means stature at the lower end of the physiological range, whereas to be stunted is a pathological state, a failure to realize the child's genetic growth potential. It is also now generally accepted that ethnic or racial differences have very little effect on the growth potential of the young child compared with the much more important effects of the environment [2].

Natural History of Stunting

The process of becoming stunted, i.e. stunting, starts very early in post-natal life, perhaps at about 3 months of age. This can be seen in figure 1, which shows the results of longitudinal studies done by three Units of the

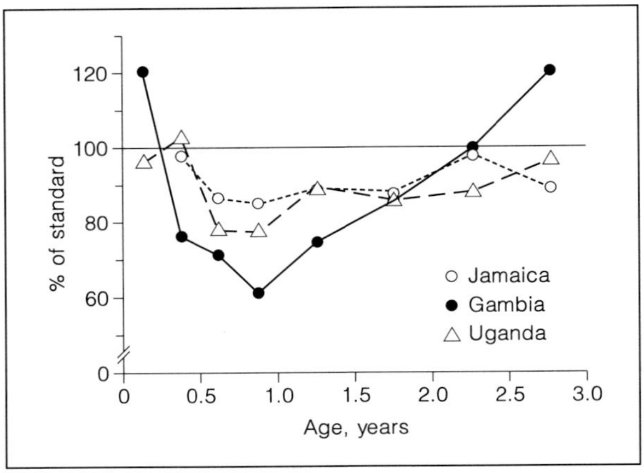

Fig. 1. Velocity curves for growth in height of pre-school children from 3 developing countries expressed as percent of Harvard Standard.

British Medical Research Council some 30 years ago, at a time when virtually all studies in third-world countries were cross-sectional [3]. The horizontal line represents the normal growth velocity; in all three countries the growth rate, initially very good, falls below this line between 3 and 6 months. A trough is reached at 12–18 months and then growth picks up, so that by 3 years a normal rate of gain is restored. This, of course, leaves the child well below the reference in its attained height. Thereafter, a common pattern is for the curve of attained height to run parallel to the reference but below it, often by as much as 15 cm or more.

The Possibility of Catch-up

The question whether catch-up of this deficit can occur has been much discussed. It has been suggested that children are 'programmed' by the age of about 2 years, so that thereafter they remain permanently stunted. This is certainly not true; the physiological capacity for catch-up is preserved, even if it is not expressed because adverse conditions continue. Clinicians have long been familiar with the remarkable catch-up that occurs when an underlying disease is treated, such as coeliac disease, hypothyroidism, or more recently Trichuris dysentery [4]. There is also good evidence of catch-up from studies on immigrants, e.g. Korean children brought to the USA [5] or Pakistani children to Sweden [6]. To my mind the most compelling example of catch-

Table 1. Height gains between 5 and 20 years: Hyderabad [9] and Guatemala [10]

	Height at 5 years % of reference	Gain, 5–20 years cm
Hyderabad		
Affluent	95	63.1
Poor	81	68.5
Guatemala		
Affluent	94	68.9
Poor	86	63.8
USA	100	66.4

up comes from a study made many years ago by Graham and his coworkers in Peru [7]. Marasmic children were admitted to hospital with gross stunting (–3.5 SD) and were still severely stunted on discharge. At the age of 4 years one group of these children was adopted by well-to-do foster-parents, and by the age of 9 had almost caught up, to be only 0.5 SD below the reference median. In contrast, those who had remained with their own very poor families by 9 years were still –2 SD below the reference.

It has been contended that in stunted children bone age, as judged by the development of the ossification centres in the wrist, is less retarded than height age [8]. It was postulated that this would lead to premature fusion of the epiphyses, so that no further growth would be possible. However, this seems not to be so; bone age in the wrist is no predictor of ultimate height. The onset of the growth spurt at puberty is delayed, but it goes on for longer.

Two patterns of later growth have been described. Table 1 shows observations from India and Guatemala [9, 10]. The stunted children never caught up, but the remarkable thing is that although they continued to live in a deprived environment their total growth between 5 and 18 years was equal to that of British or American children. In contrast, table 2 shows a pattern observed in Kenya [11]. The initially stunted children showed a tremendous growth spurt at puberty, so that having started far behind they have achieved normal height by 18 years. Although I have not been able to document it, I always had the impression during my time in Jamaica that stunted children had an increased growth spurt at puberty, so that they ended up as the magnificent athletes with which we are familiar. One explanation for this difference might be that Kenyan children are less deprived than those in India or Guatemala. Another might be that although young children have virtually

Table 2. Height gains between 10 and 18 years: Kenya [11]

	Male		Female	
	well nourished cm	malnourished cm	well nourished cm	malnourished cm
Age				
10 years	143	130	144	124
18 years	164	165	157	157
Gain	21	35	13	34

the same growth potential regardless of race, there might still be ethnic differences in the capacity for growth at puberty. The absence of a pubertal growth spurt in pygmies, together with the lack of an increase in serum IGF-1 that normally occurs at this time, would fit with this hypothesis [12]. Clearly, however, there is a problem here that needs further investigation.

Stunted Families

Some time ago we set up a study in Thailand to try to get a better understanding of the early development of the stunting process [12]. Infants were selected at birth into 3 groups according to the stature of their nearest two sibs: both normal; both stunted (below –2 SD of the reference); or one normal and one stunted. This procedure, of course, was the opposite of random selection. Its purpose was to ensure as far as possible that we would be able to make a comparison between stunted and non-stunted children. The babies were followed for 2 years and at the end of that time were classified into tertiles of height. There was a good correspondence between this classification and the one made at birth (table 3). Thus, from the stature of the sibs one could predict at birth the extent to which the infant was going to become stunted. Moreover, the infants who were destined to become most stunted not only had shorter sibs but also shorter mothers, and they themselves were slightly shorter at birth. We can therefore think of stunting as a familial process; our findings resemble those of Cole et al. [14] on the children of the unemployed in Britain. I suspect that this is a very common situation, of which the Thai study is a particular example. The question therefore arises, whether these differences between families were the result of normal genetic variation, or whether they were the result of environmental effects operating

Table 3. Height status at 2 years of Khon Kaen children classified at birth according to the height of their sibs [13]

	Group A	Group B	Group C
Sibs	N, N	N, S or S, N	S, S
Index child, length at birth, cm	50.3	49.3	48.5
Mother's height, cm	156	152	151
Index children, stunted at 2 years			
n	5/26	12/29	10/16
%	19	41	62
SD score for height at 2 years	–0.84	–1.72	–2.64

N = Not stunted; S = stunted.

on the mother, perhaps since her own childhood, and hence on the fetus in *utero*. However, the difference in length at birth was small, only 1.8 cm: the birth length of the shortest tertile at 2 years was on the 20th centile of the reference, that of the tallest tertile on the median. This does not suggest any serious malnutrition *in utero*.

I am inclined to think that this was a *homogeneous* population, within which there are the usual genetic differences; however, because of the adverse environment, the whole population is shifted to the left, compared to the stature of well-to-do Thai children, which is the same as that of the reference. The idea that this was a homogeneous population is supported by the fact that although dietary studies were made by recall – inevitably not very accurate – and blood samples were taken at 1 and 2 years, which included measurements of IGF-1 and thyroxin, we could not detect any convincing differences between the more-stunted and the less-stunted children [13].

Stunting and Nutrition

It is not enough to say that stunting is the result of a deprived environment; at the cellular level deprivation must be operating through nutritional and/or hormonal influences. There have been a number of studies in the past showing an effect of milk on linear growth. In the slump in the 1930s farmers in the UK could not dispose of their milk – people could not afford to buy it – and Boyd Orr, later to become the founding Director of FAO, suggested that it

Table 4. Effects of protein supplementation on growth and skeletal maturation of New Guinean schoolchildren [18]

	Age-adjusted increments		
	group I	group II	group III
Protein supplement, g/day	0	10	20
Height gain, cm	1.75	3.23	3.45
Weight gain, kg	1.33	1.98	2.94
Bone maturity score	18	26	28
Gain in skeletal age	0.41	0.62	0.64
Triceps skinfold gain	0.71	0.14	0.01

should be distributed free to schoolchildren under controlled conditions. The groups receiving the milk showed a small increase in height, but more remarkable was an increase in general activity; the teachers reported that the children were more difficult to control [15]. Aykroyd and Krishnan [16, 17] made similar trials in India. The study by Malcolm and coworkers in New Guinea (table 4) is particularly convincing [18]. Another important study was that of Rutishauser and Whitehead [19] in Uganda which showed that children of the Karamoja tribe, who live on meat and milk, were tall and thin, whereas those of the Buganda, whose staple is plantains, were short and fat.

So what is it in milk that promotes growth in height? Recently, Allen [20] has summarized the literature about the effects on linear growth of supplementing the diet with various nutrients, such as zinc, iron and vitamin A. The results are conflicting and certainly not impressive. She concluded that stunting is probably the result of multiple rather than single deficiencies of growth-limiting nutrients. This conclusion was supported by the findings of the US Nutrition Collaborative Research Support Programme (CRSP) carried out in infants, toddlers and schoolchildren in three countries – Mexico, Kenya and Egypt. This was a longitudinal observational study which did not include intervention, and particular attention was paid to infectious disease. Faltering in linear growth began soon after birth and mean length fell to –2 SD units by 18 months, after which there was some recovery. Thus, stunting began well before the peak incidence of diarrhoea, and started to improve while diarrhoea was still at its height. The conclusion that diarrhoea is not an important cause of long-term deficit in height is in agreement with evidence that I have summarized elsewhere [21].

Dietary analyses in the CRSP studies showed adequate intakes, according to current standards, of energy, protein and essential amino acids – lysine, tryptophan, threonine and sulphur amino acids. Nevertheless, a clear association emerged between length gain and dietary protein quality, specifically consumption of animal protein. The children with low intakes of animal products also had low intakes of fat, riboflavin, vitamins A, B_{12}, C and E. Nearly all had inadequate intakes of zinc and a large proportion were anaemic. There was also a weak but significant negative correlation between height at 30 months and intake of phytate.

It is surprising that so little attention has been given to calcium. In our study in Thailand [13], the average intake of calcium was estimated at about 140 mg/day in the first year, compared with a supposed requirement of 525 mg/day [22]; during the second year, when the requirement is supposed to be 350 mg/day the average intake was about 110 mg/day. In the tallest group of children, although the intake was far below the requirement, it was significantly greater than that of the shorter children. Similarly, low calcium intakes were also found in the CRIS studies in Kenya and Egypt, though not in Mexico. What is lacking, however, is experimental evidence that calcium deficiency causes any reduction in growth of the long bones, as opposed to poor mineralization.

The CRSP studies leave unanswered the question: what is it in milk that seems to have such a clear effect on linear growth? The authors suggest a multiple deficiency. I find it an attractive hypothesis that there might be a specific deficiency of sulphur amino acids, limiting the sulphation of proteoglycans in cartilage [23]. The CRSP data, that sulphur amino acid intakes were 'adequate', seem to rule this out, but I think that this conclusion can be countered on two grounds: first, classical estimates of the amino acid requirements of young children are based on a factorial approach, in which the adequacy of the growth component is assessed by gain in weight. This ignores the possibility that there may be special requirements for growth in length, as opposed to weight. Second, the 'requirement' is used as a cut-off point, below which an intake is inadequate, but this is a very crude approach. A much more sensitive approach is regression analysis, which in the CRSP studies showed a positive relationship between height growth and animal protein intake.

Endocrine Effects on the Growth of Long Bones

It is very difficult for a nutritionist to master the complex field of hormonal effects on the growth of long bones. I will summarize what little I know, based mainly on the work of Nilsson et al. [24], of Dauncey et al. [25] and of Pell and Bates [26], with emphasis on links to nutrition: Growth hormone (GH) is the only hormone that stimulates longitudinal bone growth in a dose-dependent fashion [24]. Energy intake does not affect GH secretion; there is no parallelism between food intake and GH levels in plasma. In fact, in malnourished children GH levels are often high [21]. There is evidence that the GH receptor is developmentally regulated and is absent in the newborn [24]. Production of the GH receptor mRNA in liver is sensitive to the intake of food [25]. There is a positive correlation between plasma T_3 and liver deiodinase activity on the one-hand and GH receptor mRNA in the liver [25]. GH stimulates the production of insulin-like growth factor IGF-1 mRNA in the liver and probably regulates the expression of IGF-1 mRNA in most tissues. The concentrations of IGF-1 and IGF-binding protein-3, which is the carrier of IGF-1, are reduced on a low food intake. The IGF-1 level apears to be affected by deficiencies of both energy and protein [27]. GH and IGF-1 have different target cells in the growth plate [24].

Thus, the action of these two hormones appears to be synergistic and thyroid hormones also plays a part. There are several points at which their production or action appears to be affected by food intake.

The work of Karlberg et al. [28] may help to link these mainly experimental observations with what is actually observed in the field. He postulates that the normal growth curve can be divided into three phases: the infant phase, which goes on for about 6 months after birth, and for which the driving force is not known; the childhood phase, which takes over at about 6 months, in which growth is driven by growth hormone; and the adolescent phase, in which the growth spurt is determined by the sex hormones. Karlberg's studies on children in Pakistan have led to the suggestion that the stunting developing in the first year results from a delayed takeover of the childhood phase; this would imply a qualitative rather than a quantitative defect. We can reconcile this idea with the fact that malnourished stunted children have high GH levels if we suppose that the defect is not in the production of GH itself but in that of its receptors.

Conclusion

The work cited here suggests that stunting results from a diet poor in quality, although it is still not clear exactly what factor or factors are limiting. Policy-makers will ask: does it matter to be stunted? and does it matter exactly what causes it? There has been heated debate on these questions, and I cannot do better than quote the words of Beaton [29]: 'We must focus on the process of becoming small and on the environmental factors – biological and social – that are the determinants of smallness.' 'Small people may cope with their environment and in that sense achieve at least the minimal health required for population survival and growth – but it has been at an important cost – a reduction in the potential for change.' 'By itself, food may not be enough to change the situation; without food, other interventions are likely to fail.'

References

1 Walterlow JC: Note on the assessment and classification of protein-energy malnutrition in children. Lancet 1973;ii:87–89.
2 Martorell R: Child growth retardation: A discussion of its causes and of its relationship to health; in Blaxter KL, Waterlow JC (eds): Nutritional Adaptation in Man. London, Libbey, 1985, pp 13–30.
3 Waterlow JC: Nutrition and Growth; in Waterlow JC (ed): Protein-Energy Malnutrition. Sevenoaks, Edward Arnold, 1992, p 198.
4 Cooper ES, Bundy TAP, MacDonald TT, Golden MHN: Growth suppression in the *Trichuris* dysentery syndrome. Eur J Clin Nutr 1990;44:285–291.
5 Lien NM, Meyer KK, Winick M: Early malnutrition and 'late' adoption: A study of their effects on Korean orphans adopted into American families. Am J Clin Nutr 1977;30:1734–1739.
6 Proos LA, Hofvander Y, Tuvemo B: Menarcheal age and growth pattern of Indian girls adopted in Sweden. II. Catch-up growth and final height. Indian J Prediatr 1991;58:105–114.
7 Graham GG, Adrianzen B: Late 'catch-up' growth after severe infantile malnutrition. Johns Hopkins Med J 1972;131:204–211.
8 Martorell R, Yarbrough C, Klein RE, Lechtig A: Malnutrition, body size and skeletal maturation: Interrelationships and implications for catch-up growth. Human Biol 1979;51:371–389.
9 Satyanarayana K, Prasanna Krishna T, Narasinga Rao BS: Effect of early childhood malnutrition and child labour on growth and adult nutritional status of rural Indian boys around Hyderabad. Hum Nutr Clin Nutr 1986;40C:131–139.
10 Martorell R, Rivera J, Kaplowitz H: Consequences of stunting in early childhood for adult body size in rural Guatemala. Ann Nestle 1990;48:85–92.
11 Kulin HE, Bwibo N, Mutie D, Santner SJ: The effect of chronic childhood malnutrition on pubertal growth and development. Am J Clin Nutr 1982;36:527–536.
12 Merimee TJ, Zapf J, Hewlett B, Cavalli-Sforza LL: Insulin-like growth factors in pygmies: The role of puberty in determining final stature. N Engl J Med 1987;316:906–911.
13 Chusilp K, Somnasang P, Kirdpon W, Wongkham S, Sribonlue P, Mahaverawat U, Yongvanit P, Sawakontha S, Waterlow JC: Observations on the development of stunting in children of the Khon Kaen region of Thailand. Eur J Clin Nutr 1992;46:475–487.
14 Cole TJ, Donnet ML, Stanfield JP: Unemployment, birthweight and growth in the first year. Arch Dis Childh 1983;58:717–721.
15 Leighton G, Clark ML: Milk consumption and the growth of schoolchildren. Lancet 1929;ii:40–43.
16 Aykroyd WR, Krishnan BG: The effect of skimmed milk, soya bean and other foods in supplementing typical Indian diets. Ind J Med Res 1937;24:1093–1115.
17 Aykroyd WR, Krishnan BG: A further experiment on the value of calcium lactate for Indian children. Ind J Med Res 1939;27:409–412.
18 Lampl M, Johnston FE, Malcolm LA: The effects of protein supplementation on the growth and skeletal maturation of New Guinean school children. Ann Hum Biol 1978;5:219–227.
19 Rutishauser RHE, Whitehead RG: Nutritional status assessed by biological tests. Br J Nutr 1969; 23:1–13.
20 Allen LH: Nutritional influences on linear growth: A general review. Eur J Clin Nutr 1994; 48(suppl 1): S75–S89.
21 Waterlow, JC: Protein-Energy Malnutrition. Sevenoaks, Edward Arnold, 1992.
22 Department of Health: Dietary Reference Values for Food Energy and Nutrients for the United Kingdom: Report on health and social subjects, No 41. London, HM Stationary Office, 1994.

23 Golden MHN: Is complete catch-up possible for malnourished children? Eur J Clin Nutr 1994; 48(suppl 1):S58–S71.
24 Nilsson A, Ohlsson C, Isaksson OGP, Lindahl A, Isgaard J: Hormonal regulation of longitudinal bone growth. Eur J Clin Nutr 1994;48(suppl 1):S150–S160.
25 Dauncey MJ, Burton KA, White P, Harrison AP, Gilmour RS, Duchamp C, Cattaneo D: Nutritional regulation of growth hormone receptor gene. FASEB J 1994;8:81–88.
26 Pell JM, Bates PC: The nutritional regulation of growth hormone action. Nutr Res Rev 1990;3: 163–192.
27 Isley WL, Underwood LE, Clemmons DR: Dietary components that regulate serum somatomedin-C concentrations in humans. J Clin Invest 1983;71:175–182.
28 Karlberg J, Jalil F, Lam B, Low L, Yeung CY: Linear growth retardation in the three phases of growth. Eur J Clin Nutr 1994;8:S25–S44.
29 Beaton GH: Small but healthy? Are we asking the right question? Eur J Clin Nutr 1989;43:863–874.

Prof. Dr. J. C. Waterlow, Department of Public Health & Policy, London School of Hygiene & Tropical Medicine, Keppel Street, London WC1E 7HT (UK)

Porrini M, Walter P (eds): Nutrition in Pregnancy and Growth.
Bibl Nutr Dieta. Basel, Karger, 1996, No 53, pp 55–59

Children on Adult Diets

Fima Lifshitz

Department of Pediatrics, Maimonides Medical Center, Brooklyn, N.Y., USA

Introduction

Five of the ten most frequent causes of death in the Western hemisphere are due to coronary heart disease (CHD), stroke, diabetes, atherosclerosis, and certain types of cancers [1]. Because these diseases are nutritionally related, diet modification in childhood has been widely recommended for disease prevention [2, 3]. The institutions that have issued similar recommendations include the American Cancer Society [4], the US Department of Agriculture (USDA), Health and Human Services [5], the American Heart Association [6], and the American Academy of Pediatrics [7]. The expert panels recommend reduction of the intake of fat and cholesterol to <30% of the total calories consumed [8]. In accordance with these recommendations, dietary habits have already changed in North America over the past two decades [9, 10]. The reduction in the consumption of milk and ice cream was 20%, butter 30%, animal fats and oils 12%, and egg 28%. A population survey in 1988 revealed that about 60% of Americans avoided high-fat and high cholesterol foods, while 57% avoided sugar altogether, including desserts and sweets.

Although dietary changes in adults may reduce the risk of morbidity and mortality [11–14], it should be pointed out that changes in life style such as reduced smoking and increased physical activity have resulted in a more significant decline in major risk factors associated with CHD than have dietary modifications [9].

Children on Adult Diets: Is It Harmful?

Children who attempt to zealously follow low-fat, low-cholesterol, or low-calorie diets may be faced with growth failure due to inappropriate nutritional intake [15, 16]. Although all experts agree that low-fat, low-cholesterol diets are not indicated below 2 years of age [7], children are dieting and modifying their nutritional intake without supervision. We have shown that 38% of high-school children were dieting to lose weight at any given time, and that 72% of them had attempted to diet in the preceding 1–2 months [17].

American Diabetic Association has shown in a recent survey that 83% of children believe in avoiding 'high-fat foods' and 71% of them worry about fat and cholesterol [18]. However, their dietary restrictions are often unsupervised, resulting in energy or micronutrient deficiencies.

Although the prevalance of nutritional growth failure due to dieting is not known, weight for height deficits have been reported in over one third of students in middle-class high school [19]. Approximately 2% of them demonstrated growth patterns suggestive of nutritional dwarfing [19].

Nonorganic nutritional dwarfing (ND) seems to be the result of the social and cultural ideal of thinness probably influenced by medical recommendations and marketing pressures rather than specific psychological behavioral disturbances [20]. Some ND patients describe a specific fear of obesity [15] or a fear of the consequences of hypercholesterolemia [21]. However, for the majority, suboptimal nutrition resulting from dietary restrictions is due to a desire to consume low-fat, low-cholesterol diets and to avoid obesity [22]. Lack of knowledge or misinformation about appropriate nutrition has also been blamed as a contributing factor [23].

As a result, children voluntarily ingest about two thirds of the calories required, and frequently skip meals. The fat ingested accounts for less than one third of the calories consumed. They avoid whole milk, red meat, oils and other fats, and most often eggs. However, avoidance of so-called 'junk food' may not provide a diet with sufficient energy, when substituted with low-calorie snacks such as raw vegetables. Additionally, it is difficult to meet the micronutrient needs of children with food that is rich in complex carbohydrates because it is bulky and filling. The consequence may be ingestion of insufficient energy and minerals leading to retardation of growth and development [17, 21, 22].

It must be recognized that the current recommendations to provide <30% of calories from fat require elimination or severe reduction of the foods which are familiar and popular to children. This may increase the risk for noncompliance and for inducing nutrient deficits and nutritional growth failure [17, 21, 24]. Also, restrictive diets for children may interfere with the development of a healthy attitude toward food and eating. Additionally, there may be iatrogenic consequences of labeling healthy children as having a medical condition requiring specific dietary treatment. Most importantly, there is a potential for increased mortality rates due to cancer and/or injury/accident-related deaths which appear to be associated with ingestion of low-fat, low-cholesterol diet [25, 26].

Children on Adult Diets: Is It Healthful?

In a longitudinal population study, a higher CHD death rate was associated with very high serum cholesterol levels (>95th percentile, >293 mg/dl) [25]. There is also good evidence that reduction of excess body weight and hypercholesterolemia in men reduce the incidence of CHD [27–29]. However, in the above-mentioned study [25], differences in mortality from CHD were minimal for individuals with serum cholesterol levels between the 20th and 85th percentiles (180–245 mg/dl), with a very small progressive increase in the relative risk from 1.3 to 2.2. Moreover, total mortality rates of the population also increased when cholesterol levels were below the 10th percentile, and the differences among individuals with serum cholesterol in the normal ranges appeared to be clinically insignificant. Therefore, the validity of intensive efforts to drastically alter dietary intake starting in early life should be questioned.

One should also question the efficiency of diets in lowering serum cholesterol. The step 1 diet in healthy adults led to a net reduction in serum cholesterol of only 0–4% [30]. More intensive diets reduced serum cholesterol concentrations by 13% over 5 years in selected high-risk men, by 6–15% in hospital outpatients, and by only 12–15% in institutionalized patients who had vigorous enforcement of their dietary intake. These trials cast doubts on the long-term efficiency of dietary intervention in reducing cholesterol levels significantly, particularly in free-living individuals. More restrictive diets combined with drug therapy would imply a commitment to many decades of treatment with methods whose safety, feasibility, and acceptability for long-term use in children have not been examined.

The last question to be asked is the effectiveness of lowering serum cholesterol concentrations in reducing mortality rates in primary prevention trials [26]. The meta analysis of six primary prevention trials of cholesterol reduction, involving 119,000 patient-years studied, revealed that the total mortality of the treatment group was not different from controls. Although mortality from CHD tended to be lower in men receiving intervention, the treatment group also had increased mortality rates due to high prevalence of cancer, accidents, suicides, or other types of violent deaths.

At present, there are no data in children on long-term dietary compliance or on the effects on serum cholesterol levels and/or efficacy in preventing CHD with dietary intervention starting in childhood. On the other hand, low-fat, low-cholesterol diets may have potential harmful effects including growth impairment, nutrient deficiencies, psychological and iatrogenic effects. Also, these diets may be associated with increased mortality due to other causes, i.e. violent deaths which are more important for children than CHD. Therefore, adult standards should not be enforced to children.

Conclusion

Children are not 'little adults' and their dietary guidelines should differ from those recommended for older persons [24]. Recommendations for children should emphasize consumptioin of a wide variety of foods since taste preferences are being developed. Guidelines should emphasize moderation and not include arbitrary numerical targets. If specific recommendations are required, then a desirable range of 30–35% of calories from fat with special emphasis given to reduced saturated fat intake may be recommended. Preferably, dietary recommendations should focus on reducing saturated fat intake to 8–12% of total energy intake.

In conclusion, children should not be given severely restrictive diets until more data are gathered.

References

1 By G: Surgeon General takes aim at saturated fats. Science, 1988;241:651.
2 Consensus Conference: Lowering blood cholesterol to prevent heart disease. JAMA 1985;253: 2081–2086.
3 Study Group of the European Atherosclerosis Society: Strategies for the prevention of coronary heart disease: A policy statement of the European Atherosclerosis Society. Eur Heart J 1987;8: 77–88.
4 American Cancer Society: Nutrition and Cancer: Cause and Prevention. New York, American Cancer Society, 1984.
5 US Department of Agriculture, US Department of Health and Human Services: Nutrition and Your Health: Dietary Guidelines for Americans. Home and Garden Bulletin No 232. Washington, Goverment Printing Office, 1985.
6 American Heart Association, Nutrition Committee: Dietary guidelines for healthy American adults. Circulation 1986;74:1465A–1468A.
7 American Academy of Pediatrics, Committee on Nutrition: Prudent life-style for children: Dietary fat and cholesterol. Pediatrics 1986;78:521–525.
8 Expert Panel: Report of the national cholesterol education program expert panel on detection, evaluation, and treatment of high blood cholesterol in adults. Arch Intern Med 1988;148:36–69, and more recently in children and adolescents. Pediatrics 1992;89:525–584.
9 Wolinsky H: Talking heart. Science 1980;21:6–10.
10 National Research Council Board on Agriculture: Designing Foods: Animal Products in the Marketplace. Washington, National Academy Press 1988, pp 18–44.
11 Multiple Risk Factor Intervention Trial Research Group: Multiple risk factor intervention trial: Risk factor changes and mortality results. JAMA 1982;248:1465–1477.

12 Lipid Research Clinics: The Lipid Research Clinics coronary primary prevention trial results: Reduction in incidence of coronary heart disease JAMA 1984;251:351–364.
13 Lipid Research Clinics: The Lipid-Research Clinics coronary primary prevention trial results: The relationship of reduction in incidence of coronary heart disease in cholesterol lowering. JAMA 1984;251:365.
14 Anderson KM, Castelli WP, Levy D: Cholesterol and mortality, 30 years of follow-up from the Framingham study 1987;257:2176–2180.
15 Pugliese MT, Lifshitz F, Grad G, Marks-Katz M: Fear of obesity: A cause of short stature and delayed puberty. N Engl J Med 1983;309:513–518.
16 Lifshitz F, Moses N, Cervantes C, Ginsberg L: Nutritional dwarfing in adolescents. Semin Adolesc Med 1987;3:255–256.
17 Pugliese MT, Weyman-Daum M, Moses N, Lifshitz F: Parental beliefs as a cause of non-organic failure to thrive. Padiatrics 1987;80:179–182.
18 Gallup Organization: Gallup Survey of Children's Eating Habits. A report prepared for the international Food Information Council/American Dietetic Association by the Gallup Organization, Princeton, 1991.
19 Pugliese MT, Recker B, Lifshitz F: A survey to determine the prevalence of abnormal growth patterns in adolescents from a suburban school district. J Adolesc Health Care 1988;9:181–187.
20 Sandberg D, Smith M, Fornari V, Lifshitz F: Nutritional dwarfing: Is it a consequence of distubed psychological functioning? Pediatrics 1991;88:926–933.
21 Lifshitz F, Moses N: Growth failure: A complication of dietary treatment of hypercholesterolemia. Am J Dis Child 1989;143:537–542.
22 Lifshitz F, Moses N: Nutritional dwarfing: Growth, dieting, and fear of obesity. J Am Coll Nutr 1988;7:368–376.
23 Dwyer JT, Feldman JJ, Mayer J: Nutritional literacy of high school students. J Nutr Educ 1970;2: 59–66.
24 Lifshitz F, Moses N, Lifshitz J: Children's Nutrition. Boston, Jones & Bartlett, 1991.
25 Martin M, Hulley SB, Browner NS, Kuller LH, Wentworth D: Serum cholesterol, blood pressure, and mortality: Implications from a cohort of 361,662 men. Lancet 1986;ii:933–936.
26 Muldoon MF, Manuck SB, Mathews KA: Lowering cholesterol concentrations and mortality: A quantitative review of primary prevention trials. Br Med J 1990;301:309–313.
27 National Academy of Sciences: Diet and Health. Implications for Reducing Chronic Disease Risk. Washington, National Academy Press, 1989.
28 Dorr AE, Gundersen D, Schneider JC, Spencer TW, Martin WB: Cholestipol hydrochloride in hypercholesterolemic patients: Effect on serum cholesterol and mortality. J Chron Dis 1978;31:5–14.
29 Dayton S, Pearce ML, Hashmoto S, Dixon WJ, Tomyasu U: A controlled clinical trial of a diet high in unsaturated fat in preventing complications of atherosclerosis. Circulation 1969; 39–40(suppl II):1–63.
30 Ramsay L, Yeo WW, Jackson PR: Dietary reduction of serum cholesterol concentrations: Time to think again. Br Med J 1991;303:953–957.

Dr. F. Lifshitz, Department of Pediatrics, Maimonides Medical Center,
4802 Tenth Avenue, Brooklyn, NY 11219 (USA)

Porrini M, Walter P (eds): Nutrition in Pregnancy and Growth.
Bibl Nutr Dieta. Basel, Karger, 1996, No 53, pp 60–73

Growth in Italian Children

Ivan Nicoletti[a], *Silvano Milani*[b], *Anna Bossi*[c]

[a] Centro Studi Auxologici, Firenze,
[b] Statistica Medica e Biometria, Università di Pisa, e
[c] Istituto di Biometria e Statistica Medica, Università di Milano, Italia

The first investigation on the growth of Italian children, cited in the literature as one of the earliest to apply the criteria of modern auxology, is Pagliani's in 1879 [1]. Thereafter, many decades elapsed before Italian auxological research papers appeared with some frequency and only after World War II was this field of study fully developed.

The most important publications up to the early 1960s are reported in De Toni's book [2] while many of the later ones are referred to in an article by Nicoletti and Pelissero [3] and in two volumes edited by Nicoletti [4, 5].

The aim of this paper is to summarize some of the main characteristics concerning the height and biological maturation of Italian children, based on information in the Italian scientific literature of the last 15–20 years and on as yet unpublished research on population samples from two north-Italian regions. These characteristics will then be compared with those of subjects from other European populations. We shall often be referring to the Tuscan population, both because most of our studies deal with that population and because the anthropometric characteristics of Tuscan subjects are similar to those of most populations in the center-north of Italy.

Height of Italians at End of Growth: Regional Differences and Secular Trend

Table 1 shows the height means of Italian conscripts. It can be seen that the height varies from region to region. Each row refers to a region of Italy, proceeding from the north through the center and south to the islands.

Table 1. Height means of Italian conscripts during the past century

		Subjects examined in		Subjects born in				
		1879/83	1894	1931	1941	1951	1961	1967
1	Piemonte	164.9	164.6	169.7	170.8	172.3	174.2	173.8
2	Valle D'Aosta	–	–	168.7	170.7	172.0	174.4	174.2
3	Lombardia	165.3	164.7	169.1	169.9	171.3	173.8	174.1
4	Trentino Alto Adige	–	–	169.4	171.0	172.8	175.1	175.9
5	Veneto	166.6	166.7	170.1	171.0	171.4	179.9	175.4
6	Friuli Ven. Giulia	–	–	171.4	173.0	174.0	176.2	176.7
7	Liguria	165.5	165.1	169.8	171.2	171.8	174.0	174.2
8	Emilia Romagna	165.3	165.3	169.6	170.6	172.0	174.6	174.8
9	Toscana	165.6	164.9	170.1	171.0	172.5	175.3	174.9
10	Umbria	164.2	163.9	167.5	168.1	171.3	173.4	174.1
11	Marche	163.8	163.8	167.7	168.9	171.0	173.5	174.0
12	Lazio	164.3	164.0	168.0	170.3	170.7	173.8	174.2
13	Abruzzo	163.2	162.5	165.8	167.0	169.1	172.0	172.6
14	Molise	–	–	164.4	165.9	168.0	170.8	172.5
15	Campania	163.5	162.9	165.1	165.7	168.3	170.9	172.3
16	Puglia	163.5	162.2	164.7	166.1	168.3	171.3	172.4
17	Basilicata	162.6	159.3	162.6	163.7	166.3	169.7	171.4
18	Calabria	163.1	161.1	163.8	165.0	167.3	169.5	171.6
19	Sicilia	163.5	161.7	164.9	165.7	167.6	170.4	171.1
20	Sardegna	161.9	159.9	163.6	165.2	166.2	169.0	169.9
North-center		–	–	169.3	170.5	171.7	174.3	174.5
South and islands		–	–	164.7	165.7	167.8	170.6	171.6
Italy		164.2	163.3	167.4	168.7	170.0	172.6	173.6

Data from Guerci A: Contributo alla conoscenza della dinamica costituzionalistica della popolazione italiana. Studio diacronico sull'incremento staturale dei giovani in età di leva dal 1879 al 1969. Arch Antrop Etnol 1977;107:305–315; and from Istat: Annuario 1990; 75–76.

An examination of the secular trend for height in the various regions reveals that the differences between the center-north and south and islands have declined, with mean values of height in the south and the islands tending to approach those in the center-north.

Terrenato and Ulizzi [6], who analyzed data for subjects born in 1874, 1903, 1918, 1938, 1951, 1954, 1957 and 1960, found that after a long period of more or less constant annual increments, there was a clear-cut acceleration for those born in the decade following the end of World War II and a deceleration after 1954, more marked in the north than in the south. Available data show that for subjects born in 1960, the increment was close to zero in Tuscany and Liguria, whereas in southern regions the increment exceeded 0.1% per year [6].

Some results from studies in the early 1980s led Terrenato and Ulizzi [6] to conclude that, compared to some 20 years earlier, all regions displayed a narrower range of adult male heights and the regional means were progressively approaching the national one. These findings suggest the following interpretation: the improvement in social, economic and cultural conditions as well as the decrease in the differences among regions have reduced the variability of environmental factors. Consequently, the secular trend tends to vanish since the threshold for the genetic potential's complete expression is reached.

Does the increase in height of males over the past century apply to females as well? We have no long-term information on Italian women, lacking the data base used for men – compulsory medical screening for military service. However, surveys of the school-age population in recent decades point to an affirmative answer.

A recent study of ours (fig. 1) has pointed out that over the last 15–20 years, the mean height of males in Tuscany has remained constant whereas that of females has increased.

Thus, the secular trend in Tuscany appears to have approached zero for males while it continues to be positive for females. We are at a loss to explain this discrepancy; the only hypothesis that comes to mind is that the trend starts later among girls than boys, but even if that is the case, why is it so?

Height during Growth

The mean height of boys and girls between 7 and 15 years of age in various parts of Italy [3] shows the regional differences already noted for males at the end of the growth period.

A critical examination of the literature indicates that in childhood there is fair agreement between the heights of Tuscan youngsters and those of children and adolescents from other regions of central and northern Italy [3]. But Tuscan children are definitely taller than their counterparts in southern and insular regions, though – according to recently published data for some southern regions – these differences are not detectable, or only barely so, in the early years of life [4, 5].

The discrepancy is greater between Tuscan and Sardinian subjects than between children from Tuscany and either Campania or Puglia; but the data for these last two regions are more recent [5].

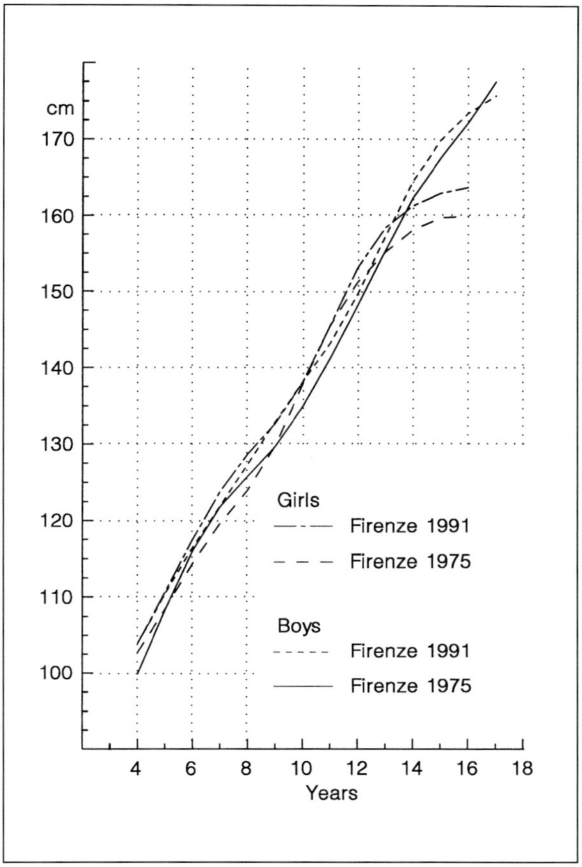

Fig. 1. Mean height of boys and girls in Tuscany in 1975 and 1991.

The following two points emerge: (1) at birth and in the first few years of life, differences in height between one region and another are negligible from a practical point of view; these findings support current views on the relative roles of environmental and genetic factors in the growth process: the former clearly predominates at the beginning of life; (2) after the third year of life up to adulthood, the differences are such that the 50th percentile for southern and insular populations approximates the 25th percentile of the standard of central and northern populations [4].

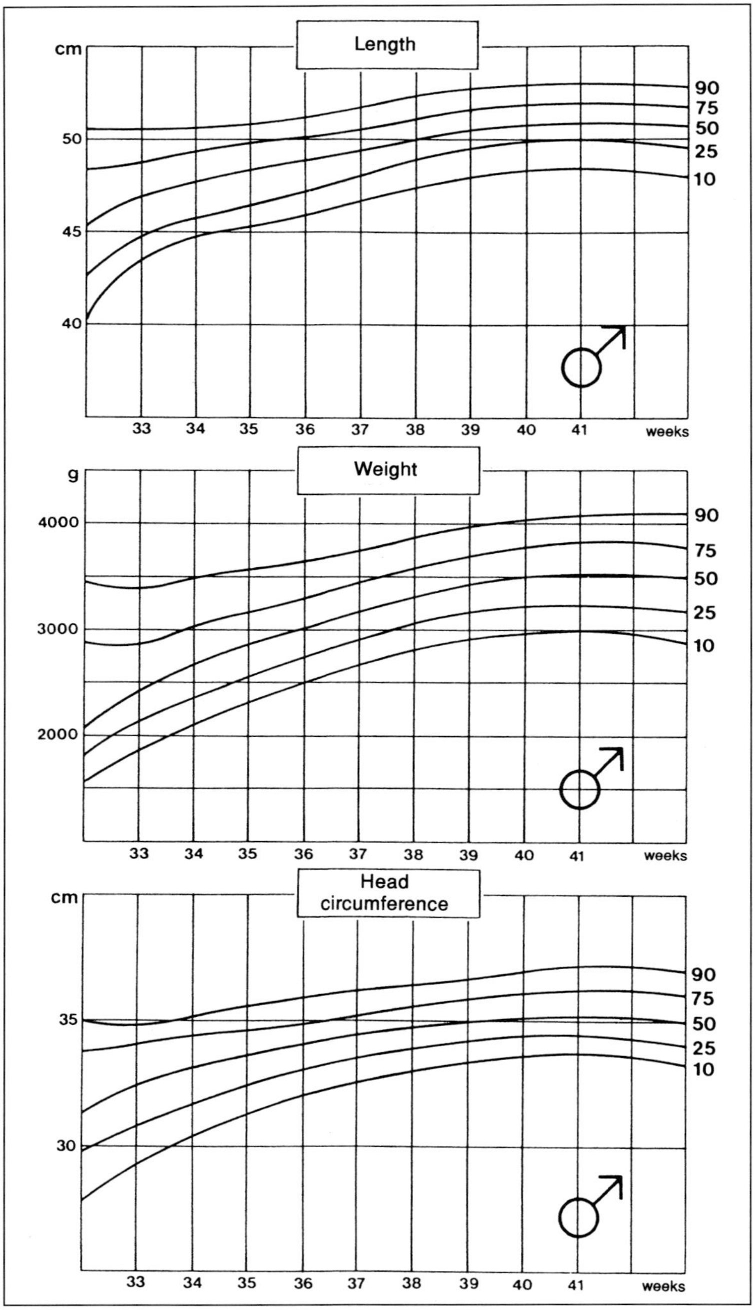

Fig. 2. Standard for neonates, males [7].

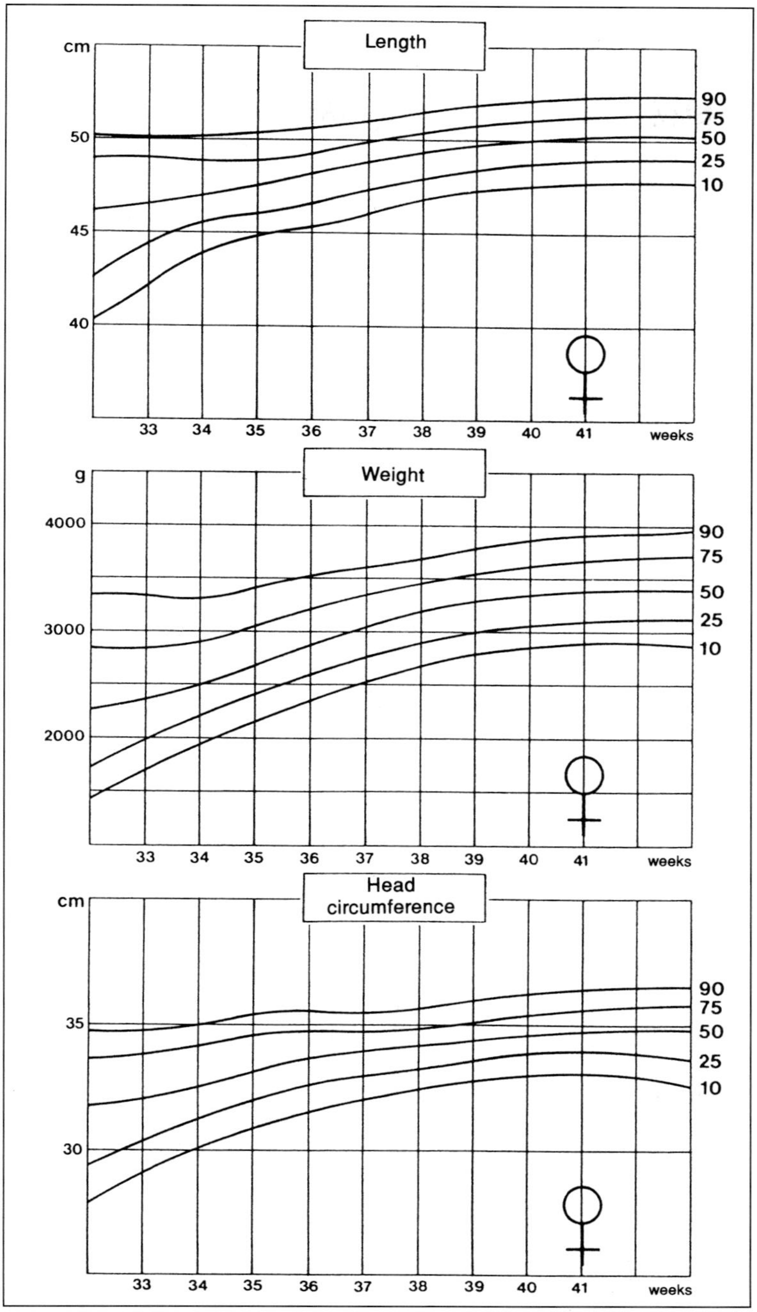

Fig. 3. Standard for neonates, females [7].

Italian Height Standards

Thus, the following height standards are necessary and sufficient for the needs of Italian medical practitioners and nutritionists: (1) one standard for neonates (fig. 2, 3), valid for the whole country [7]; (2) one standard for infants up to the age of 3 (fig. 4, 5; adapted from Milani et al. [8]) valid for the whole country; (3) one standard for 3- to 18-year-olds valid for the center-north regions (fig. 6, 7) [4]; (4) an analogous one for the south and the islands (not yet established).

Comparisons with Height Standards of Other Countries

We undertook a comparison with standards of other countries, fitting by the so-called JPPS model [9] the mean heights for our population of Tuscan children and adolescents, and those of some European populations. The standards we examined were the British, French, Swedish, Hungarian and Czechoslovakian [5].

For the entire period under consideration, the Tuscan curve is above the British one. For males, the differences are slight at the younger ages and at the end of growth, being more substantial (about 3 cm) during puberty; the situation for females is analogous, but with smaller differences at puberty. More specifically, the difference in height between the two male populations is less than 10 mm in the early years, increases to 30–35 mm for 12.5- to 14-year olds, and thereafter declines to something like 10 mm; the heights of the girls, instead, differ by about 10 mm up to around 11 years of age when the difference rises to about 15 mm and then vanishes after the age of 13.

We will not go into the details of the comparison with other height standards in order to concentrate on another significant distinction between the Tuscan population and the other European groups we have studied: namely the age at peak of pubertal growth velocity.

The age at which growth velocity peaks in the British population is 13.5 years for boys and 11.5 for girls. These ages are somewhat higher than the corresponding figures for Tuscany – 13.1 and 10.9 years, respectively, while the difference between the sexes is very similar in both populations – 2 years for the British, slightly more than that for the Tuscans.

Comparable or still larger differences in the age of maximum growth velocity emerge from comparisons with other west-European populations.

Populations tend to diversify at puberty even if their growth curves are superimposable in childhood and pre-puberty. Comparisons with Hungarian and Czechoslovakian standards provide further support for this conclusion

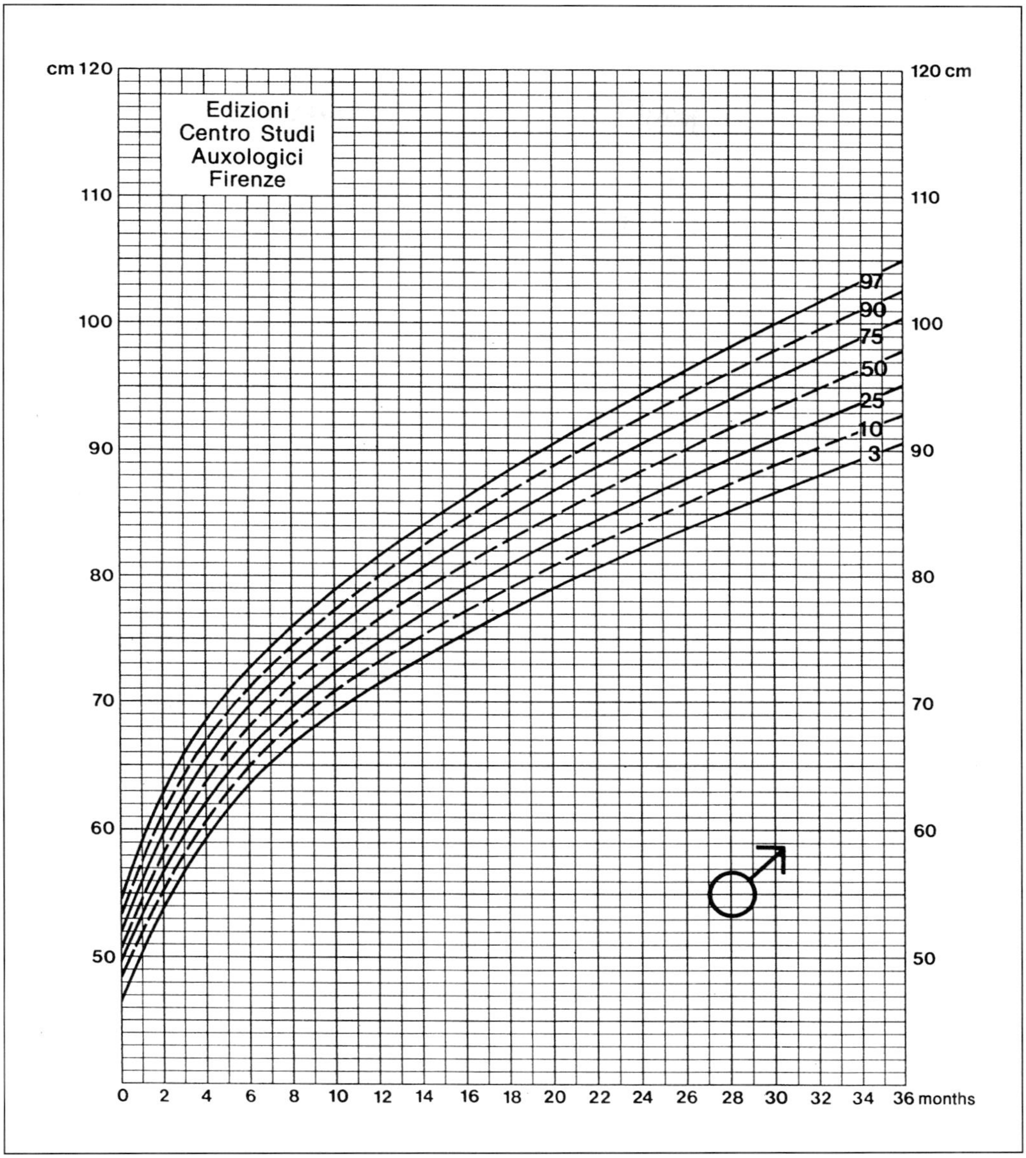

Fig. 4. Standard for length of infants up to the age of 3, males. Adapted from Milani et al. [8].

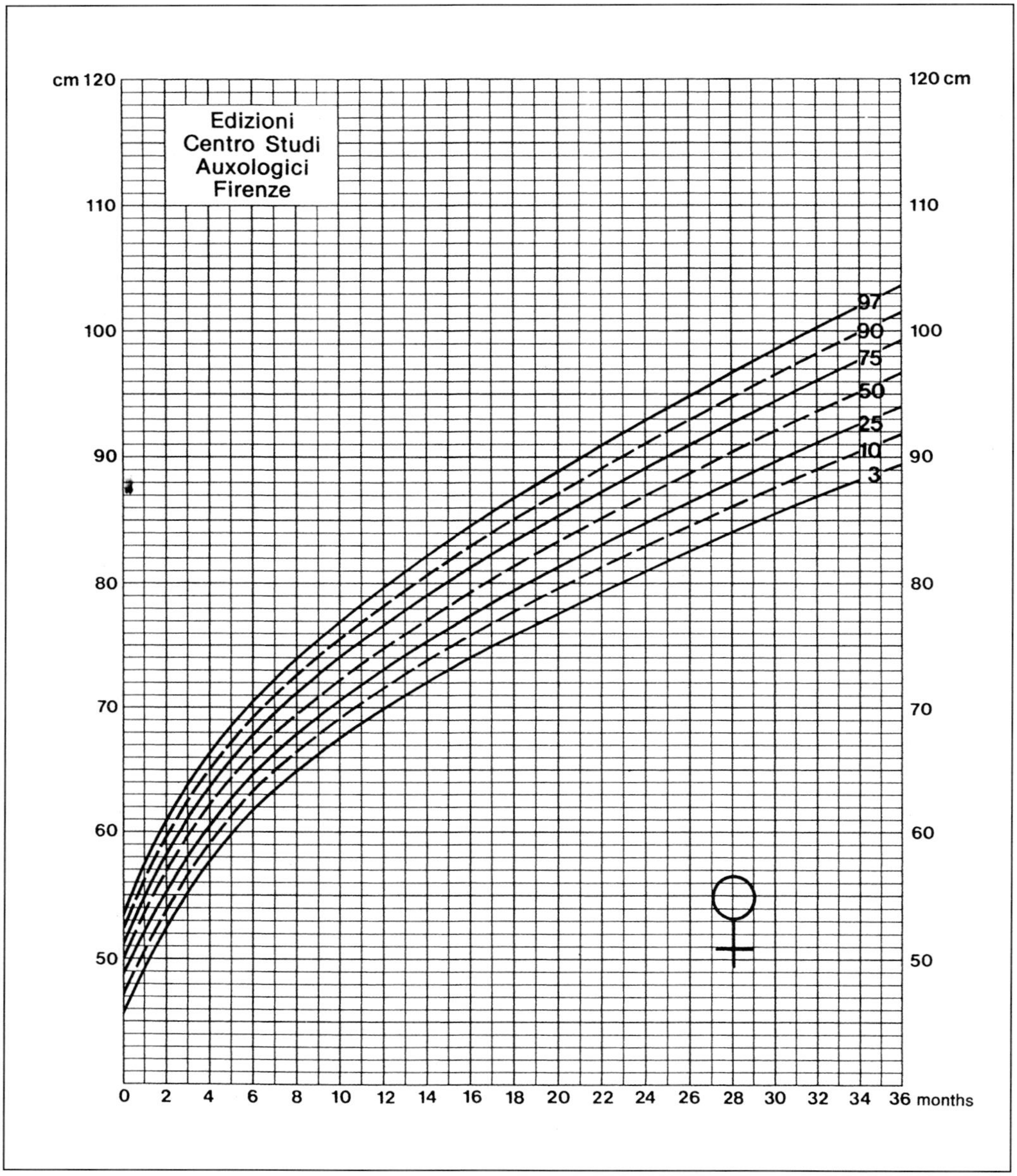

Fig. 5. Standard for length of infants up to the age of 3, females. Adapted from Milani et al. [8].

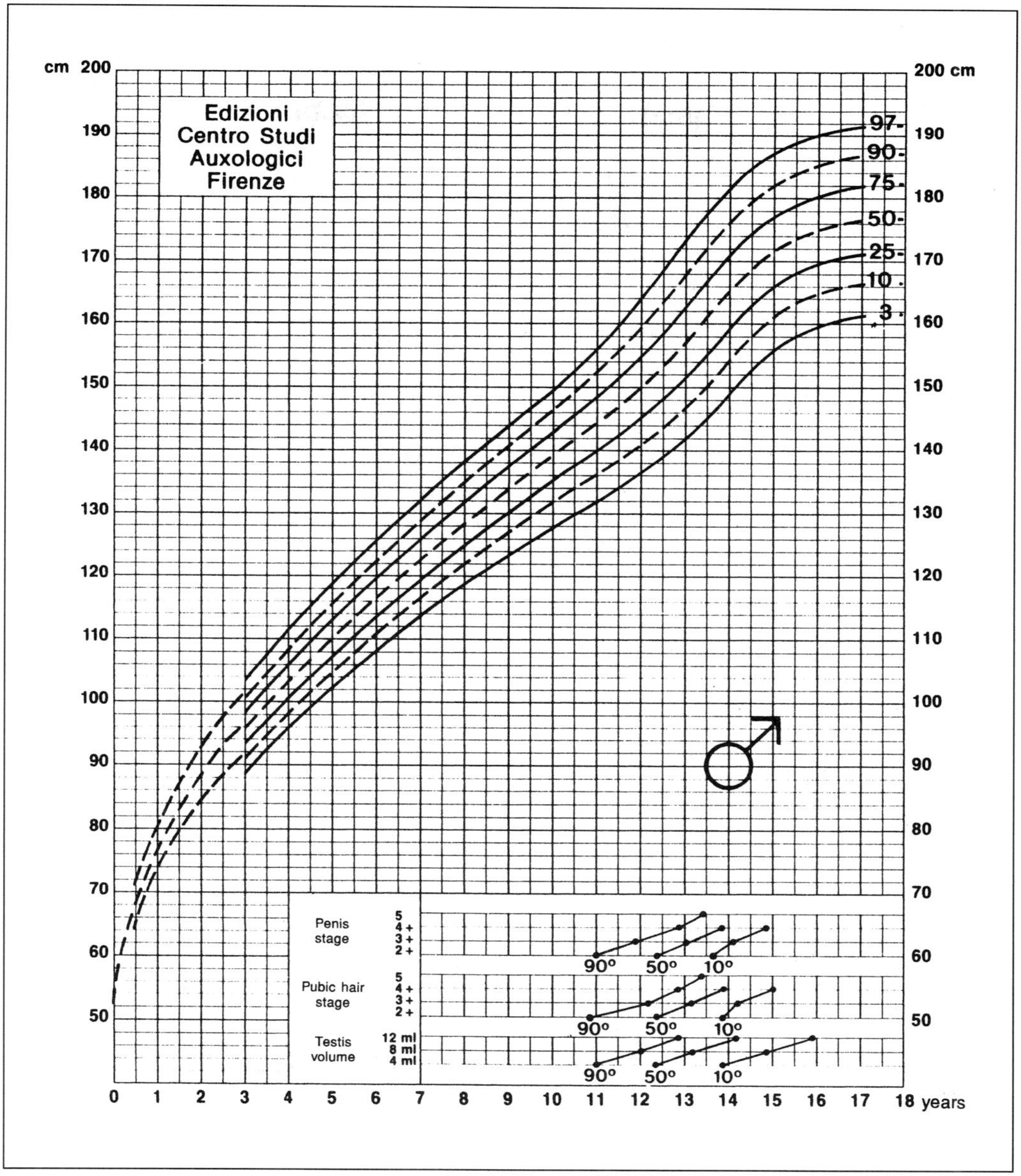

Fig. 6. Standard for height of 3- to 18-year-old males for the center-north regions.

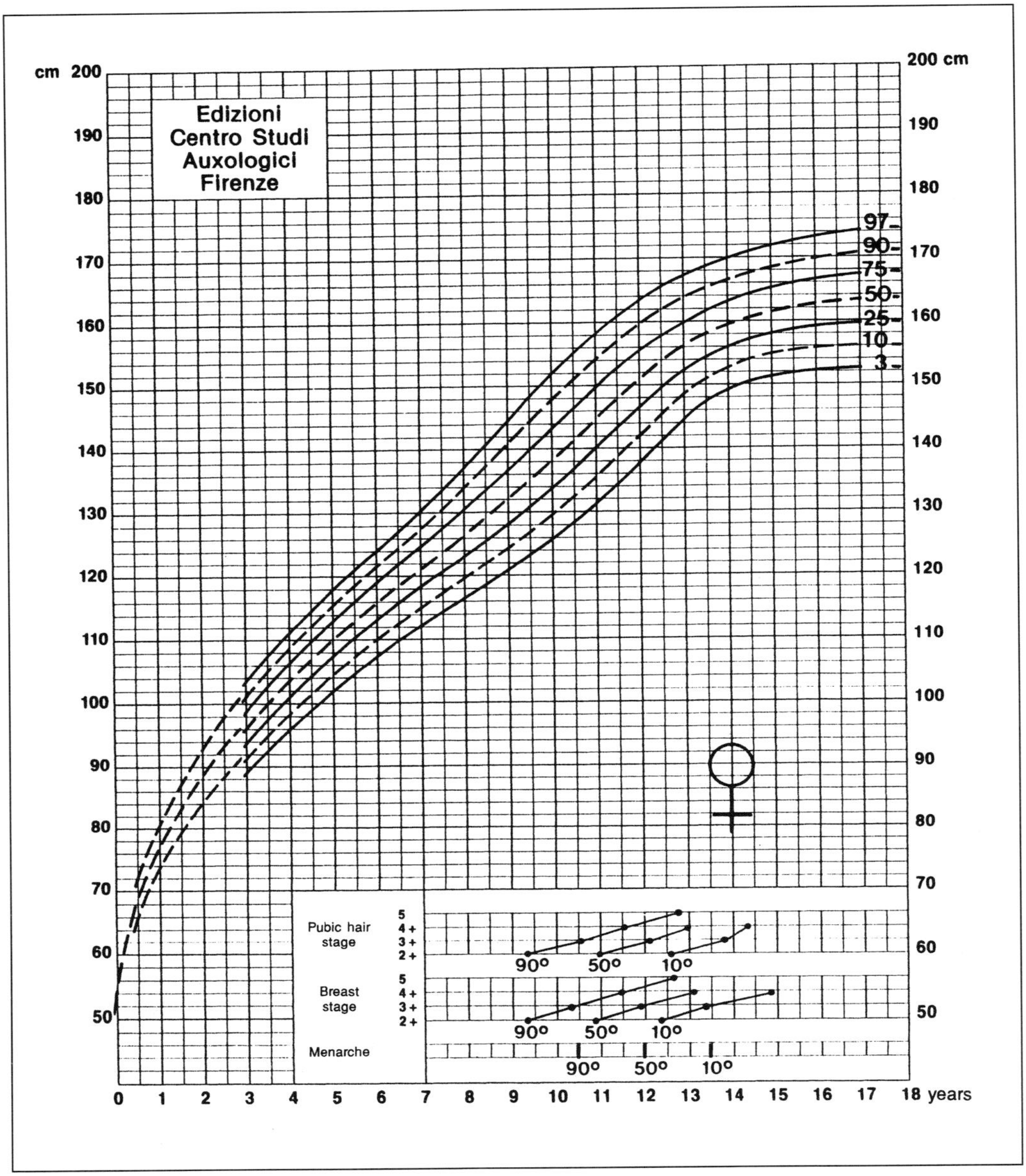

Fig. 7. Standard for height of 3- to 18-year-old females for the center-north regions.

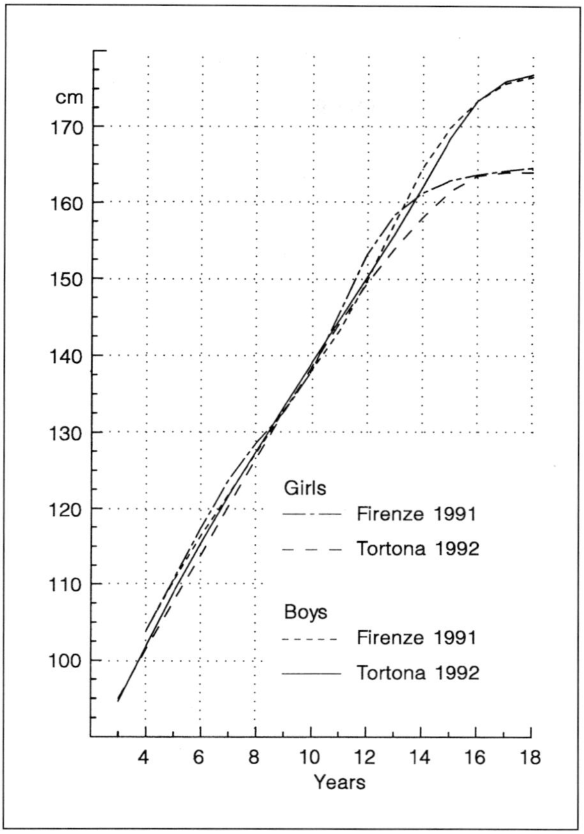

Fig. 8. Height curves of boys and girls from Tuscany (Firenze) and Piedmont (Tortona).

[5]. A recent comparison of data from Tuscany and Piedmont (fig. 8) shows that even between geographically close regions, differences at puberty are seen. We suggest that genetic factors become more prominent at puberty, accentuating ethnic differences.

Growth and Social Class

According to some studies, the influence of certain environmental conditions, such as those grouped under the term 'social class' is considerable up to the age of 7, after which, at least in developed countries like Great Britain, it becomes negligible [10]. For the age range spanning from 3 to 6 months up to 2 years, growth means provide a particularly sensitive index of a population's or subpopulation's socio-economic conditions [11]. When, in a developing country, height at the age of 2 years starts to approach the standards of developed populations, it indicates an improvement in the socioeconomic status of that country's population.

We have no detailed studies on the relationship between height and social class in today's Italy. But we presume that the smaller difference in mean heights among Italian regions is due, at least in part, to improved socioeconomic conditions among the lower classes in the south and the islands.

Biological Maturation of Italian Children

Our discussion about 'maturation' will focus on the age of menarche and the appearance of secondary sexual characteristics as well as on skeletal maturation.

In Italy today the age of menarche falls between 12 and 12.8 years, depending on the author and on the region [5]. In our recent research we found the mean age to be 12.25 in Tuscany and 12.55 in Piedmont.

The mean for Italian regions is lower than that in north- and central-European countries as well as in some east-European ones like Hungary and Russia, which is around 13 years, whereas it is similar to that for Greece, found to be about 12.5 years, and similar also to the means for Spain, the United States, Argentina, Brazil and Venezuela [5].

As in other European countries, in Italy too the age of menarche has advanced by 2 years. This finding emerged from the comparison of Pagliani's mean for the city of Turin in 1879 and that noted somewhat more than a century later by Benso et al. [12].

Our research shows that the development of sexual characteristics of males and females occurs in Tuscany a little earlier than in Piedmont and earlier today than in the past.

As for skeletal maturation, we found that it is more advanced in Tuscans than in British subjects at every age after 6.5 years; at lower ages the differences are negligible [13]. Analogous findings have since been reported for other Italian regions by other researchers [13].

Concluding Considerations

In all populations, environmental factors tend to prevail in influencing growth during the prenatal and early postnatal periods. Thereafter, the influence of genetic factors tends to prevail and, as suggested by our studies on Italian children, tends to reach its maximum expression at puberty.

Italian boys and girls reach sexual and skeletal maturity earlier than several other continental European populations.

In Italy, regional differences in mean height are tending to diminish as a result, most likely, of the improvement in economic and social conditions in the poorer regions of the south and the islands.

In Tuscany, the secular trend of the height of males is tending to vanish, but the analogous trend of females is still positive.

References

1 Tanner JM: A History of the Study of Human Growth. Cambridge, Cambridge University Press, 1981, pp 198–200.

2 De Toni G, De Toni E, Zunin C: Auxologia. Torino, Edizioni Minerva Medica, 1965.

3 Nicoletti I, Pelissero G: Considerazioni su uno standard di crescita staturale per la popolazione italiana. Acta Med Auxol 1979;11:117–126.

4 Nicoletti I: La crescita del bambino italiano. Firenze, Edizioni Centro Studi Auxologici, 1992.

5 Nicoletti I, Pelissero G: Auxologia epidemiologica e sociale; in Nicoletti I (ed): Auxologia Normale e Patologica. Firenze, Edizioni Centro Studi Auxologici, 1994, pp 611–632.

6 Terrenato L, Ulizzi L: Genotype-environment relationships: An analysis of stature distribution curves during the last century in Italy. Ann Hum Biol 1983;10:335–346.

7 Milani S, Bossi A, Cortinovis I, Bertino E, Fabris C, Mostert M, Benso L: Neonatal anthropometric standards: An update. Acta Med Auxol 1990;22:27–45.

8 Milani S, Cortinovis I, Bossi A: Longitudinal growth standards for length, weight and head-circumference of Italian babies up to three years. Fifth International Auxology Congress, Exter, 1988, abstr 49.

9 Jolicoeur P, Pontier G, Pernin MO, Sempé, M: Logiciel JPPS, modélisation de la courbe de croissance statural de l'enfant. Madrid, Laboratoires Serono SA, 1989.

10 Lasker GW, Mascie-Taylor CGN: Effects of social class differences and social mobility on growth in height, weight and body mass index in a British cohort. Ann Hum Biol 1989;16:1–8.

11 Beaton GH: Nutrition research in human biology: Changing perspectives and interpretations. Am J Hum Biol 1992;4:159–177.

12 Benso L, Lorenzino C, Pastorin L, Barotto M, Signorile F, Mostert M: The distribution of age at menarche in a random series of Turin girls followed longitudinally. Ann Hum Biol 1989;16:549–552.

13 Nicoletti I: Crescita e maturazione scheletrica. Firenze, Edizioni Centro Studi Auxologici, 1990.

Ivan Nicoletti, Piazza Madonna degli Aldobrandini, 1, I–50123 Firenze (Italy)

Porrini M, Walter P (eds): Nutrition in Pregnancy and Growth.
Bibl Nutr Dieta. Basel, Karger, 1996, No 53, pp 74–89

The WHO Growth Chart: Historical Considerations and Current Scientific Issues

Mercedes de Onis[a], *Ray Yip*[b]

[a] Nutrition Unit, World Health Organization, Geneva, Switzerland;
[b] Division of Nutrition, Centers for Disease Control, Atlanta, Ga., USA

Introduction

Growth assessment is the single measurement that best defines the health and nutritional status of children, just as it provides an indirect measurement of the quality of life of an entire population. Building from a history of pediatric practice, the premise for the use of growth as an indicator of health and nutritional status lies in the fact that 'poor' growth performance reflects, for the majority of the children who are so designated, deviations from favorable environmental conditions that support optimal growth and development in childhood.

The designation of a child as having impaired growth implies some means of comparison with a 'reference' child of the same age and sex. This need has made the choice of a normative growth reference an important issue that has received considerable attention in the last decades; especially as growth retardation during early childhood and its adverse implications for child development remains a disturbing problem worldwide [1]. This paper reviews the background and history of the currently used NCHS/WHO growth charts, discusses some of the contemporary scientific issues and outlines the recommendations made by a WHO Expert Committee held in November 1993 [2] related to the revision and development of a new international growth reference.

Background and History

For many years the pioneer data for children developed in the 1940s by Meredith [3] at Iowa were used mostly for individual assessment. These data were derived from a small and unrepresentative sample of US children, most of whom were of a high socioeconomic status. In the 1960s and 1970s and following on from the Iowa data, two data sets were frequently used as growth references: the Harvard growth curves [4] and the Tanner growth curves from the UK [5]. In 1966, a simplified combined-sexes version of the Harvard growth curves was widely disseminated by the World Health Organization [6] and effectively established the Harvard curves as the international growth reference. This data, derived from the growth of Caucasian children in Boston from 1930 to 1956, offered the advantages of having been compiled longitudinally, of being widely available in Nelson's Textbook of Pediatrics [4], and of being used already by pediatricians in many countries. The major criticisms of the Harvard data come from the small numbers of children included, the limited genetic representativeness, and out-of-datedness.

During the next decade, as a result of the limitations of the Harvard and other available reference data and the desire in the US for a more contemporary reference that better represented the population of that country [7], the US National Academy of Sciences recommended in 1974 a new set of anthropometric data as the US national growth reference [8, 9]. A combined National Center for Health Statistics (NCHS) and Centers for Disease Control (CDC) task force, using cross-sectional data from the US Health Examination Surveys of NCHS and longitudinal data from the Fels Research Institute, constructed a set of smoothed percentile distributions for attained weight, height, and head circumference from birth to 18 years [10, 11]. A description of these newer charts and their limitations can be found elsewhere [12, 13].

In 1975, a working group was convened to advise WHO on the use of anthropometric indicators of nutritional status in surveys and for nutritional surveillance. As a result, recommendations were made on the use of children's height and weight data which included the use of a reference population for international comparison [14]. Among other criteria, the sample was to be from a cross-sectional study of a well-nourished population and was to include at least 200 children of each sex in every age group. The sampling methods were to be well defined and reproducible and the anthropometric measurements were to have been made by trained staff using well-designed, calibrated equipment. The anthropometric measurements were to include a selection of measurements sufficient to evaluate the nutritional status of children. Finally, the anthropometric data were to be available for anyone wishing to use them and the procedures used for smoothing the curves were to be adequately described and documented [14].

Although none of the three anthropometric data sets available at that time (the measurements of Dutch children reported by Van Wieringen [15]; those of US National Academy of Sciences [9]; and those of British children reported by Tanner et al. [5]) met all the criteria recommended by the WHO working group, the NCHS/CDC growth reference was identified as being the most suitable for use as the international growth reference [14, 16, 17].

The NCHS/CDC growth curves were developed by combining two distinct data sets representing different age groups compiled in different decades. The reference for ages 0–23 months is based upon a group of children in the Ohio Fels Research Institute Longitudinal Study from 1929 to 1975. This reference reflects the growth of children who were fed primarily with infant formula and who were of restricted genetic, geographic, and socioeconomic background. The height curves for this part of the reference was based on recumbent length or supine measurements (the Fels curves). The reference from 2 to 18 years was based on data of three cross-sectional US representative surveys conducted from 1960 to 1975. The height curves for this part of the reference were based on standing height measurements (the NCHS curves). It was intended that length measurements be interpreted with the Fels curves, and height measurements with the NCHS curves. One major limitation of this reference is therefore that it is made up from two unrelated samples or sets of curves. Ideally, a reference should be based on a single set of curves or on a single survey sample.

In 1980, a mainframe computer software version of the reference was developed by the Centers for Disease Control to facilitate the interpretation of growth data from surveys or clinical studies. In order to formulate the software-based reference, the original height or weight distributions were slightly modified by a normalization procedure [12]. Throughout the 1980s, several microcomputer-based software versions of the NCHS-WHO reference were developed and supported by CDC and WHO [18]. These software-based references have contributed to the wide acceptance of the concept of the international reference because they simplified the handling of anthropometric data from surveys, surveillance, and clinical studies.

Issues Related to Reference Selection and Application

Reference versus Standard

On theoretical grounds it is important to distinguish between a reference and a standard [19]. A *reference* is defined as a tool for grouping and analyzing data and provides a common basis for comparing populations without making inferences about the meaning of observed differences. A *standard,* on the other hand, embraces the notion of a norm or desirable target, and thus involves a value judgement. However, because reference data embody certain characteristics or patterns of normality, they have been widely used to make inferences about the health and/or nutrition of individuals and populations; that is, they have been treated as optimum targets, or standards, and any deviations have been assumed to have a fixed and particular meaning. Much of the justification for this is provided by extensive evidence that, in populations, the effect of genetic differences on the growth of children is small in comparison with the large differences observed due to environmental factors.

Therefore, recognizing that in practice it is almost impossible to prevent the use of reference data as standards for judging the nutritional status of individuals and populations, it is recommended that care should always be taken to choose references that resemble, as far as possible, true standards, so that the same deviation from the reference data has the same biological meaning. Beyond the consideration of choosing a reference population, efforts should be concentrated on the appropriate use of the reference as a general guide for screening and monitoring, and not as an absolute criterion to define 'malnutrition' or pathology.

Local versus International Reference Data

An international reference is clearly needed to compare the nutritional status of populations in different parts of the world. There is evidence that the growth in height and weight of well-fed, healthy children, or children who experience unconstrained growth, from different ethnic backgrounds and different continents is reasonably similar at least up to 5 years of age [20, 21]. It is accepted that there is some variation in the growth patterns among children of different race or ethnic groups in developed countries; however, these variations are relatively minor compared to the large worldwide variation in growth related to health, nutrition and socioeconomic status [22, 23]. To illustrate this point using two very different populations, figure 1 shows how the growth performance of Indian children not subject to socioeconomic and dietary constrains, nearly corresponds to the NCHS/WHO international reference values. The same pattern of growth is seen among affluent children in seven different cities of India [24]. Similarly, in Brazil, children in the highest quartile of per capita income have a height-for-age Z-score distribution that overlaps with that of the international reference [25]. Figure 2 illustrates how the entire height-for-age Z-score distribution for Brazilian children shifts to the right in moving from the lowest to the highest income quartile distribution. Several other affluent populations from different ethnic backgrounds have

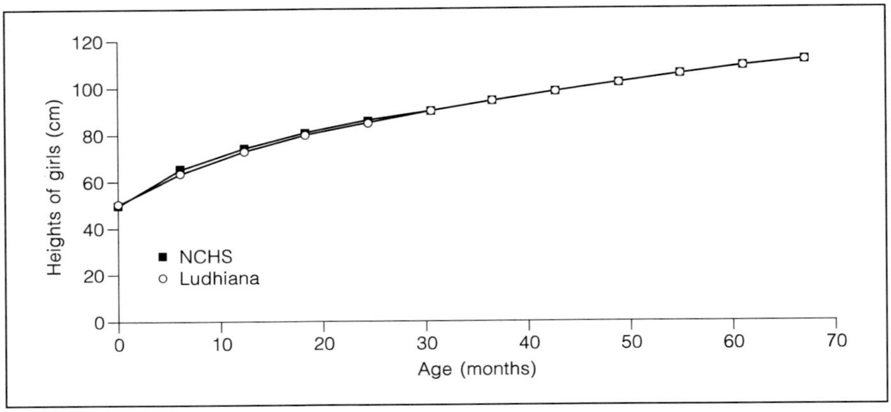

Fig. 1. Comparison of the 50th percentile of heights of affluent Indian girls (Ludhiana) with NCHS/WHO reference values. Source: Agarwal et al. [24].

been shown to have a growth pattern similar to the international reference. For this reason, the use of a common reference has the advantage of uniform application allowing international comparisons without losing the usefulness for local application. Such advantages outweigh the disadvantage of not taking into account minor racial and ethnic variations.

Beyond lacking value for international comparisons, there are also several reasons for not developing a local reference or standard [26]: (1) many populations in less-developed areas experience growth deficits as a result of poor health and nutrition, and therefore the reference developed from such populations has less screening value for the detection of health and nutritional disorders; (2) significant secular changes in growth status within a relatively short period of time may render a local reference less useful for clinical screening; (3) proper reference development is not a task that can be done easily and frequently, and (4) it is very costly to develop local references.

Probably more important than which reference is chosen is the way in which the reference is interpreted, and the clinical and public health decisions that will be based upon it. Criteria for selection of cut-off points used in the assessment of 'high-risk' individuals and populations should be based on evidence of increased risk for mortality, morbidity or impaired performance. Future research should therefore attempt to identify the range of child growth associated with optimal long-term health outcomes, as well as ranges associated with specific adverse outcomes. These results may lead to changes in the traditionally used cut-off points based on statistical criteria.

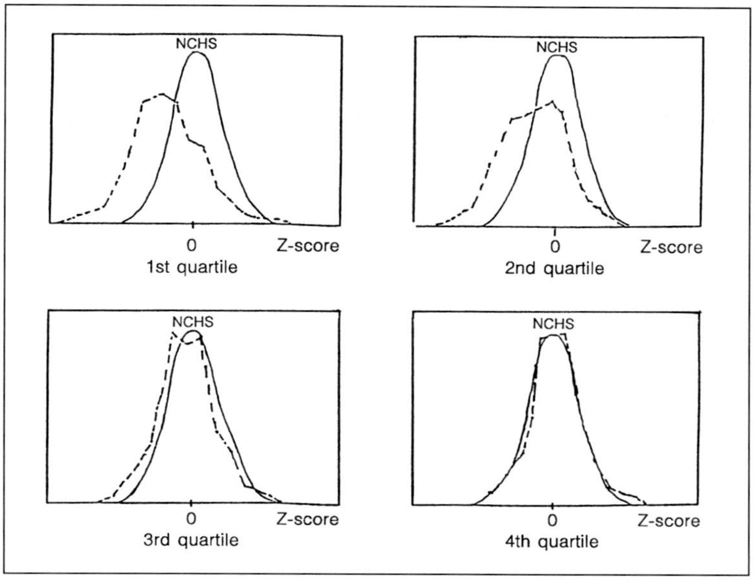

Fig. 2. Height-for-age by quartiles of per capita family income (Brazil, 1989). Source: Monteiro et al. [25].

Factors Affecting the Use and Interpretation of Growth References

A number of nonpathological factors may influence growth during childhood. Some of these factors have the potential to alter the interpretation of growth status at both the individual and population level.

Feeding Practice. Existing data in the published literature suggest that the mode of feeding during infancy significantly influences growth patterns. Breast-fed infants living under favourable conditions and studied in various geographical areas have been reported to follow negative trends relative to the NCHS-WHO weight-for-age, and possibly length-for-age, percentiles during the latter half of the recommended exclusive breastfeeding period [27, 28].

Racial and Ethnic Variation. There is evidence that the following variations exist even though relatively minor in contrast to variations related to socioeoconomic status: (1) weight-for-height status is lower for children of the Indian subcontinent [29]; (2) Hispanic children in Central and South America appear to have higher weight-for-height status [29, 30]; (3) Black children have lower birth-weight and smaller size during infancy, but exceed the size of white children after 2–3 years of age [31].

Sex. In the age range of concern (birth to 10 years) the average weights and heights of boys are consistently larger than those of girls. Gender-specific references are therefore recommended, with the possible exception of disaster situations where simplification of procedures may require the use of unisex references.

Size at Birth. Birth-size and intrauterine growth status appear to be strong determinants of later childhood growth status even in the range commonly accepted as normal birth weight. Intrauterine growth-retarded infants grow more slowly than preterm infants of the same birth-weight [32].

Parental Stature. Parental size, especially height, is a determinant of both birth-weight and later childhood growth.

Altitude. The environmental factor that has most effect on growth but is not associated with socioeconomic factors is altitude (or oxygen partial pressure); high altitude leads to reduced birth weight and lesser subsequent growth [33].

With the exception of sex and possibly of infant feeding mode, most of the factors discussed here do not seem to warrant a separate reference to account for the observed differences. Rather, awareness of these factors can help to adjust for possible variations when comparisons are made across groups.

Regardless of the approach taken, the definition of the reference population must include consideration of variability. A reference with an inappropriately small variability will result in more children classified as being of clinical concern as compared to a reference with more variability. Since references are intended for international use, they should reflect the variability observed cross-nationally in well-nourished, healthy populations, including at least partly variability due to genetic differences.

The Growth of Breast-Fed Infants

As part of the preparatory work for the recently held Expert Committee on the use and interpretation of anthropometry [2], WHO established a working group on infant growth to examine the growth of breast-fed infants living under favourable environmental conditions and see whether they differed substantially from accepted international references; and whether those differences were of practical importance to clinical practice and public health policy.

Concern had been expressed in the literature that breast-fed infants, living under favorable conditions, were growing less well than expected when compared to the NCHS-WHO growth reference. The negative deviations in weight-for-age appeared large enough to cause health workers to draw inappropriate conclusions about the growth of these infants. This possibility was particularly worrisome in environments where breast-feeding is key to survival. Alternatively, true growth failure in a breast-fed infant may have been misinterpreted

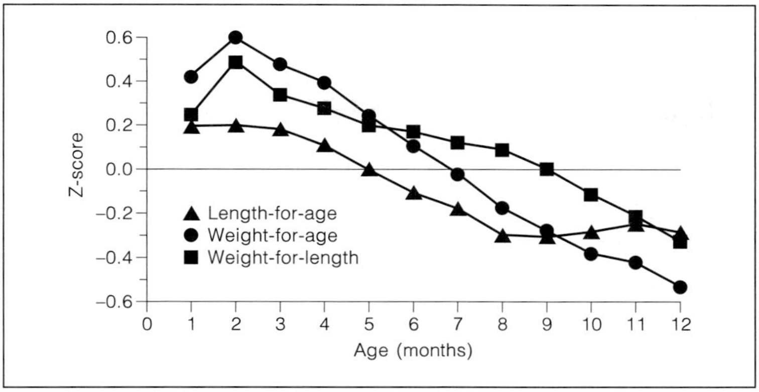

Fig. 3. Mean Z-scores of infants in the 'breast-fed set', relative to the NCHS-WHO reference [27].

as the 'normal negative deviation' in growth expected when breast-fed infants are compared to the current international reference.

A survey of investigators with data sets on growth of breast-fed infants was conducted. Seven data sets fulfilled the criteria for inclusion in the initial review and were examined in detail: one each from Canada, Denmark, Finland, Sweden, the UK, and two from the US. Of the 453 infants followed in the seven studies, 226 were breast-fed for at least 12 months and not given solid food, formula or other milks before 4 months of age. These infants were referred to as the 'breast-fed set'. Maternal education levels were high in the studies reporting this variable, and mean birth-weight in the seven studies was above 3,400 g. A complete description of these analyses is available elsewhere [27, 28].

Figure 3 presents the Z-scores patterns of the 'breast-fed set' infants relative to the NCHS/WHO reference values. Mean weight-for-age declined continuously from 2 to 12 months to a low of almost –0.6 SD at 12 months. The magnitude of the decline in length-for-age was not as great, with the mean Z-score tending to stabilize or increase after 8 months; the mean value at 12 months was approximately –0.3 SD. Mean weight-for-length at 12 months was also below the NCHS-WHO reference median (approximately –0.3 SD). The declines in Z-scores observed during the period of complementation are difficult to interpret. One possibility is that these are artifacts due to technical problems in the construction of the NCHS-WHO references [2]. The declines may also be a consequence of specific weaning practices within the population studied, or the result of other physiological effects attributable to continued

Table 1. Comparison of the data sets used in the NCHS/WHO growth reference

Data source	Representative?[1]	Age range years	Stature measurement	Study design
Fels	no	0.0–2.99	recumbent	longitudinal
NCHS	yes	2.0–17.99	standing	cross-sectional

Source: Gorstein et al. [34].
[1] Representative of the entire US population.

breast-feeding, or even due to unidentified characteristics in the populations used to construct the NCHS-WHO reference and the 'breast-fed set'.

Based on this review [27, 28], the working group concluded that infants fed according to WHO recommendations and living under conditions favoring the attainment of genetic growth potential grew less rapidly and deviated significantly from the reference. Given the short- and long-term consequences of growth failure, the dangers both of the premature introduction of complementary foods and their undue delay – described as the 'weanling's dilemma' – the working group recommended the development of a new weight and length reference which would enhance the nutritional management of infants. The specific issues to take into consideration in the development of such a reference and the characteristics of the sample to be selected as a reference population have been described elsewhere [2].

Of special concern is the role of complementary foods in determining growth patterns. Research needs to be done in advantaged populations to examine the effects of the quality of complementary foods on growth and other health outcomes from 4 months onwards. The lack of this information makes it difficult both to evaluate growth in the latter half of infancy, and to identify the etiology and measure the rates of stunting and/or wasting in infants within and between populations.

Current Problems of the NCHS/WHO International Growth Reference

Height-for-Age Disjunction

As mentioned earlier, one major limitation of the current international reference is that it is made up from two unrelated sets of curves (table 1). As a result, there is a significant disjunction in the height curve between the length-based

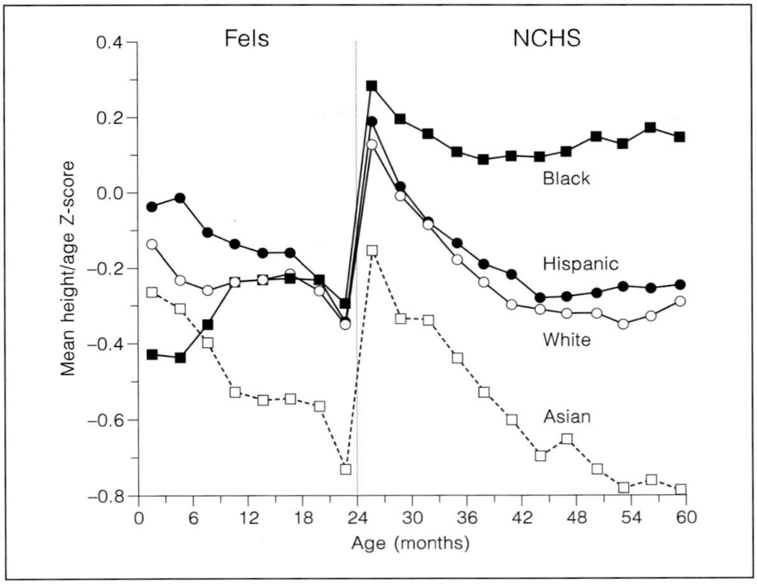

Fig. 4. Mean height-for-age Z-scores across age for four ethnic groups of low-income US children monitored by the CDC Pediatric Nutrition Surveillance System, illustrating the marked shift in mean Z-scores across the 24-month disjunction. From Yip et al. [31].

Fels curves for children under 2 years of age and the height-based NCHS curves for older children. This marked discrepancy in estimated height status immediately before and after 24 months of age, where the two sets of data merge, is responsible for the abrupt change consistently noted at 2 years of age in the mean Z-score of height-for-age and the prevalence of low height-for-age when analysing growth data from large surveys. A detailed description of the nature and magnitude of the height-for-age disjunction can be found elsewhere [2, 34].

Figure 4 uses the mean height-for-age Z-score across age for four racial groups of low-income US children monitored by CDC to illustrate the significant changes described above [31]. The impact of the 24-month disjunction on the prevalence of low height-for-age for these children is illustrated in figure 5. The magnitude of these deviations warrants caution when using the current reference to interpret the growth status of children covering a wide range of ages, as in surveys or in the context of growth monitoring.

Upward Skewness of Reference Population

The height distributions of children of a given age are normally (symmetrically) distributed. On the other hand, the distributions of weight-for-

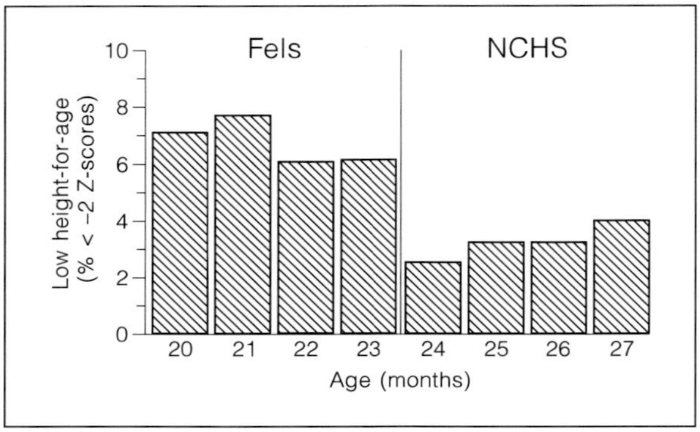

Fig. 5. Change in prevalence of low height-for-age of low-income US children monitored by the CDC Pediatric Nutrition Surveillance System, illustrating the impact of the 24-month disjunction. Source: WHO [2].

age and weight-for-height are normal in some developed countries but are skewed towards the higher end in others. In fact, the current NCHS-WHO reference based on United States' children is markedly skewed, reflecting a substantial level of childhood obesity. Using such a reference as a 'standard' for optimal growth or health is unwarranted since the upward skewness may reflect an 'unhealthy' characteristic of the sample. In addition, given the secular trends of increased childhood and adult overweight observed in the USA and in some other populations, it is of concern that, unless this issue is dealt with, future references will tend to be further and further skewed to the right, thereby resulting in the misclassification of overweight children as 'normal'.

One potential solution is to use the standard deviation of the lower half of the distribution to model the upper half. The main justification for this suggestion is that variations in the lower half of the weight distributions among developed countries (non-constrained populations) are relatively small. However, at present it is felt that not enough is known about the normality of weight distribution in children of different ages so that the issue of skewness should be the subject for further research [2].

Revision Effort of the Current NCHS-WHO Growth Reference

Because of the potential for misinterpretation of anthropometric data resulting from the above-mentioned problems of the current reference, a revision was carried out by CDC in 1990 to correct some of the irregularities. The revision added additional data from the original US national

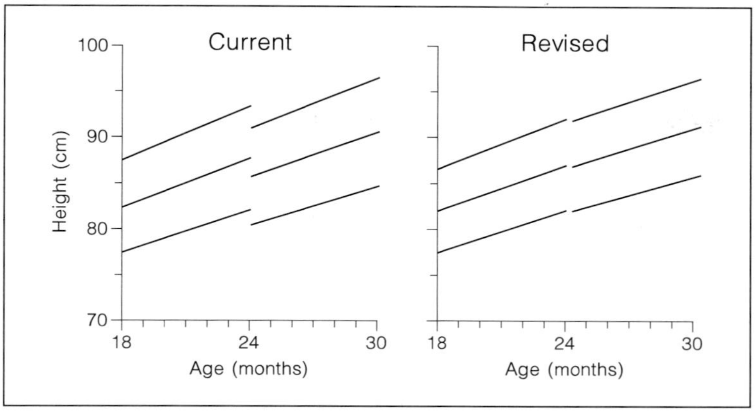

Fig. 6. Height-for-age disjunction at 24 months of the current NCHS-WHO reference and of the revised NCHS reference. Source: WHO [2].

survey to expand the sample size for more reliable curve formulation, as well as to reduce the use of the Fels sample for 12–24 months. In addition, different statistical procedures were developed to ensure greater accuracy of the formulated curve. Figure 6 shows that the gap of height values between the Fels curves and the NCHS curves was reduced from the original 2.5 cm to 0.5 cm after the revision. It should be noted that this reflects only the difference between recumbent length and standing height, which in this case was small since the children were properly stretched for the measurement.

Since a national United States representative sample of children under 12 months of age was not available, the Fels sample was retained for this age group. However, a new procedure based on modelling the individual growth pattern was used to overcome the inadequate measurement points for the sample. Unlike the original reference, where the Fels and the NCHS curves were kept as separate sets, the revised reference merged them at 12 months of age to create a single set, as at this age an adequate fit was observed.

The adoption of the revised curves as the updated international reference for general use is not recommended for several reasons [2]. Rather, the formulation of a truly international reference based on well-conducted surveys covering broad populations from several countries might be more acceptable than data from a single country. Some of the desirable characteristics for the anthropometric data to be used in the development of such a reference population are listed in table 2.

Table 2. Desirable characteristics for anthropometric data to be used in the development of an international growth reference

Several countries from different geographical regions should be included
Based on healthy populations with unconstrained growth (not necessarily representative)
Adequate sample sizes and procedures
The raw data should be available
Age range from birth to adolescence
Standardized quality control and measurements with documentation of standardization procedures
For adolescents, measures of sexual maturity should be available
Secular trends in growth should be small or absent

Adapted from: WHO [2].

Growth Velocity Curves

Growth velocity curves will detect growth faltering earlier than attained growth charts. Velocity (rate of growth) represents what is happening now, whereas attained growth represents the sum of all that has happened in the past [35]. However, such curves are more difficult to interpret and require repeated visits at regular intervals. In addition, growth over short periods of time show high variability, a finding which along with the low precision of some measurements will limit the ability to detect growth changes. For proper decision-making, it is important to use time intervals for which the expected growth is greater than the combined errors of two repeated measures.

The significance of a change in weight or height clearly depends on the time over which that change is observed. Failure to gain weight over 3 days has a different significance from failure to gain weight over 3 months. Therefore, when velocities are reported, it is important to indicate the time over which they are measured. Different time intervals will be appropriate in different situations. They will depend on, among other things, the age of the children and whether weight or height gain is the main object of interest. The smaller time intervals will have the greatest error. The effect of the wide variability in normal growth rates on the use and interpretation of growth velocity reference data is also an important research issue.

One important difficulty encountered in constructing longitudinal reference data is the sample size required for the accurate location of the extreme percentiles. This difficulty can be overcome by constructing curves the shape of which are based on longitudinal data, but whose amplitudes (in the sense of distance between the centiles) are based on a large-scale cross-sectional survey [36]. This was the method used in creating one of the longitudinal standards presently available [37].

Accurate estimates of velocity also require that the same children should be examined at defined intervals. Longitudinal studies of this kind will usually not be feasible. Even if they are, another important issue concerns the frequency of data collection. Should data be collected once a month, once every 3 months, or using any other time interval? For length and weight during infancy, intervals between 2 weeks and 3 months have been proposed [38]. In available velocity standards the data points are at intervals of 3 months or longer. If a falling off in velocity is to provide early warning of growth failure, it must be detected over much shorter periods than 3 months. The expected velocity over any given month can be obtained by interpolation, but the SD of 3-month gains cannot be applied to gains over 1 month. The shorter the interval, the larger the SD [39]. There is no doubt that clinical practice requires velocity reference data; however, considering the difficulties involved in constructing growth velocity curves, the justification for their general public health use would have to be carefully evaluated before embarking on this task.

Conclusions and Recommendations of the WHO Expert Committee Related to the Revision and Development of an International Reference

The Expert Committee on 'Physical status: the use and interpretation of anthropometry' reaffirmed the previous WHO position of using a single international reference. However, because of the significant technical drawbacks of the current NCHS-WHO reference population, especially for population-based application, an update or replacement in the near future was recommended [2]. Updating a reference, or developing an entirely new one, is an extremely complex, costly and time-consuming undertaking. Such a daunting task cannot be treated lightly. To be of lasting value, therefore, it is clear that the next reference must be exceptionally well prepared. The Expert Committee identified some of the desirable characteristics for anthropometric data to be used in the development of a new growth reference. Applying the Committee's recommendation that the reference be based on data from numerous countries will help prevent both the technical and political difficulties that have arisen from using a single country's child-growth pattern as a worldwide 'standard'.

In addition to developing a more acceptable international reference, efforts should be concentrated on the appropriate use of the reference. The reference should be used as a general guide for screening and monitoring and not as a fixed standard that can be applied in a rigid fashion to children from different ethnic, socioeconomic, and nutritional and health backgrounds.

For clinical or individual-based application, the NCHS/WHO growth curves should be used as a screening tool to detect children at greater risk of health or nutritional disorders; and they should not be viewed as a diagnostic label for 'malnutrition'.

For population-based application, the reference population can be used adequately for cross-comparison and monitoring purposes. In a given population or community, a high prevalence of anthropometric deficit will be indicative of significant health and nutritional problems; however, it is not only those children below the cut-off point who are at risk, the entire population is at risk, and the cut-off point should be used only to facilitate the application of the indicator.

References

1 de Onis M, Monterio C, Akre J, Clugston G: The worldwide magnitude of protein-energy malnutrition: An overview from the WHO Global Database on Child Growth. Bull WHO 1993;71:703–712.

2 WHO. Physical Status: The use and interpretation of anthropometry. Report of WHO Expert Committee. Technical Report Ser 854 Geneva, World Health Organization, 1995.

3 Meredith HV: A 'physical growth record' for use in elementary and high schools. Am J Pub Health 1949;39:878–885.

4 Stuart HC, Stevenson SS: Physical growth and development; in Nelson WE (ed): Textbook of Pediatrics, ed. 5. Philadelphia, Saunders, 1950, pp 14–73.

5 Tanner JM, Whitehouse RH, Takaishi M: Standards from birth to maturity for height, weight, height velocity, and weight velocity: British children 1965. Arch Dis Child 1966;41:454–471.

6 Jelliffe DB: The assessment of the nutritional status of the community. World Health Organization Monogr Ser 53. Geneva, WHO, 1966.

7 Owen GM: The assessment and recording of measurements of the growth of children: Report of a small conference. Pediatrics 1973;51:461–466.

8 Food and Nutrition Board, National Academy of Sciences–National Research Council: Comparison of body weights and lengths and heights of groups of children. Nutr Rev 1974;32:284–288.

9 US Food and Nutrition Board: Committee on Nutrition Advisory to CDC. Comparison of body weights and body heights of groups of children. US Department of Health, Education and Welfare, Public Health Service. Atlanta, Center for Disease Control, 1974.

10 National Center for Health Statistics: Growth curves for children birth–18 years, United States. Vital and health statistics, Ser 11, No 165, DHEW publ 78–1650. National Center for Health Statistics. Washington, Government Printing Office, 1977.

11 Hamill PVV, Drizd TA, Johnson CL, Reed RB, Rouche AF, Moore WM: Physical growth: National Center for Health Statistics percentiles. Am J Clin Nutr 1979;32:607–629.

12 Dibley MJ, Goldsby JB, Staehling NW, Trowbridge FL: Development of normalized curves for the international growth reference: Historical and technical considerations. Am J Clin Nutr 1987;46: 736–748.

13 Dibley MJ, Staehling N, Nieburg P, Trowbridge FL: Interpretation of Z-score anthropometric indicators derived from the international growth reference. Am J Clin Nutr 1987;46:749–762.

14 Waterlow JC, Buzina R, Keller W, Lane JM, Nichaman MZ, Tanner JM: The presentation and use of height and weight data for comparing nutritional status of groups of children under the age of 10 years. Bull WHO 1977;55:489–498.
15 Van Wieringen JC: Secular changes of growth. 1964–1966 height and weight surveys in the Netherlands. Leiden, Netherlands Institute for Preventive Medicine, TNO, 1972.
16 World Health Organization: A Growth Chart for International Use in Maternal and Child Health Care. Guidelines for Primary Health Care Personnel. Geneva, WHO, 1978.
17 World Health Organization: Measuring Change in Nutritional Status. Geneva, WHO, 1983.
18 Fichtner RR, Sullivan KM, Trowbridge FL, Carlsson B: Report of the Technical Meeting on Software for Nutrition Surveillance. Food Nutr Bull 1989;11:57–61.
19 WHO Working Group: Use and interpretation of anthropometric indicators of nutritional status. Bull WHO 1986;64:929–941.
20 Martorell R: Child growth retardation: A discussion of its causes and its relationship to health, in Blaxter KL, Waterlow JC (eds): Nutritional Adaptation in Man. London, Libbey, 1985, pp 13–30.
21 Graitcher PL, Gentry EM: Measuring children: One reference for all. Lancet 1981;ii:297–299.
22 Habicht JP, Martorell R, Yarbrough C, Malina R, Klein RE: Height and weight standards for preschool children. How relevant are ethnic differences in growth potential? Lancet 1974;i:611–615.
23 Martorell R, Mendoza F, Castillo R: Poverty and stature in children; in Waterlow JC (ed): Linear Growth Retardation in Less Developed Countries. New York, Raven Press, 1988, pp 57–70.
24 Agarwal KN, Agarwal DK, Benakappa DG, et al: Growth performance of affluent Indian children (under-fives). Growth standard for Indian children. Scientific Report 11. New Delhi, Nutrition Foundation of India, 1991.
25 Monteiro CA, Benicio MH, Gouvea NC: Growth and nutritional status of Brazilian children: Findings from the 1989 National Health and Nutrition Survey. Country Studies on Nutritional Anthropometry, Doc NUT/ANTREF/1/91. Geneva, WHO, 1991.
26 Sullivan K, Trowbridge F, Gorstein J, Pradilla A: Growth references. Lancet 1991;337:1420–1421.
27 WHO Working Group on Infant Growth. An evaluation of infant growth: The use and interpretation of anthropometry in infants. Bull WHO 1995;73:165–174.
28 WHO Working Group on Infant Growth. An evaluation of infant growth. Doc WHO/NUT/94.8. Geneva, WHO, 1994.
29 Victora CG: The association between wasting and stunting: An international perspective. J Nutr 1992;122:1105–1110.
30 Martorell R, Mendoza FS, Castillo RO, Pawson IG, Budge CC: Short and plump physique of Mexican-American children. Am J Phys Anthropol 1987;73:475–487.
31 Yip R, Scanlon K, Trowbridge FL: Trends and patterns in height and weight status of low-income US children. Crit Rev Food Sci Nutr 1993;33:409–421.
32 Binkin NJ, Yip R, Fleshood L, Trowbridge FL: Birthweight and childhood growth. Pediatrics 1988; 82:828–834.
33 Yip R, Binkin NJ, Trowbridge FL: Altitude and child growth. J Pediatrics 1988;113:486–489.
34 Gorstein J, Sullivan K, Yip R, de Onis M, Trowbridge F, Fajans P, Clugston G: Issues in the assessment of nutritional status using anthropometry. Bull WHO 1994;72:273–283.
35 Tanner JM: Fetus into Man: Physical Growth from Conception to Maturity. Cambridge, Harvard University Press, 1978.
36 Tanner JM: Use and abuse of growth standards; in Falkner F, Tanner JM (eds): Human Growth: A Comprehensive Treatise, ed 2. New York, Plenum Press, 1986, Vol III, pp 95–109.
37 Tanner JM, Whitehouse RH: Clinical longitudinal standards for height, weight, height velocity and weight velocity and the stages of puberty. Arch Dis Child 1976;51:170–179.
38 Cole TJ: The use and construction of anthropometric growth reference standards. Nutr Res Rev 1993;6:19–50.
39 Waterlow JC: Protein-Energy Malnutrition, ed 2. London, Edward Arnold, 1992, p 194.

Dr. Mercedes de Onis, World Health Organization, Nutrition Unit, 20, Av. Appia,
CH–1211 Geneva 27 (Switzerland)

Porrini M, Walter P (eds): Nutrition in Pregnancy and Growth.
Bibl Nutr Dieta. Basel, Karger, 1996, No 53, pp 90–102

Biochemical Markers of Skeletal Growth in Children

Francesco Branca

Istituto Nazionale della Nutrizione, Rome, Italy

Introduction

Recent developments of immunological and molecular biology techniques have allowed major advances in the understanding and management of growth disorders. However, surprisingly little research is being carried out on the physiopathology of failure to thrive, despite the worldwide diffusion of the phenomenon. This is partially due to the inadequacy of current methods to monitor fine changes in growth rates. New anthropometric techniques, such as knemometry, allow measurements of small length increments, but still on a time scale that encompasses several cycles of chondrocyte replication and maturation, thus smoothing out oscillations of growth. Biochemical markers may help to overcome these limitations and have become the object of considerable interest [1].

Skeletal growth is the combination of bone lengthening and bone modelling. Longitudinal growth takes place at the level of the epiphyseal plate, at a rate proportional to the number of proliferating chondrocytes per column, the average rate of division and the final cell size in the hypertrophic zone, before incorporation into bone [2]. Bone modelling is a continuous process of resorption, carried out by osteoclasts, and formation, carried out by osteoblasts, that allows the achievement of final adult shape of the growing bone. In long bones, this mechanism allows the drifting of the mid-shaft and the reduction of the metaphyseal funnel into a narrower shaft. Bone resorption and formation are 'coupled' by endocrine and paracrine factors, while the mechanisms that coordinate lengthening and modelling have not been elucidated.

Ideal markers should allow monitoring of the different components of skeletal growth with the least discomfort to the subject and early enough to allow the identification of factors inducing growth arrest or acceleration. However, established biochemical markers only allow the measurement of bone formation and resorption, while no specific indicators are yet available to monitor the activity of the epiphyseal plate. This paper will provide an update of the literature on biochemical markers of growth and will touch on some of the most promising future developments.

Cartilage Markers

Among the numerous components of cartilage matrix (collagen II and IX, proteoglycans, several proteins [3]) only few have been used as turnover markers and none as a growth marker. Due to their wide diffusion in the body, most matrix constituents would in fact be too nonspecific. The only applications are the measure of the C-propeptide of collagen II in synovial fluid and chondroitin sulfate in serum to study cartilage damage in rheumatic diseases.

A specific marker for the turnover of the epiphyseal cartilage could be collagen X, a non-fibril-forming collagen synthesised by hypertrophic chondrocytes. The molecule has been extracted from fetal human growth plate and has been demonstrated by immunolocalization in the hypertrophic and calcifying cartilage zone [4], but has not yet been identified in biological fluids. Hypertrophic chondrocytes in chicks have been shown to produce a 21,000 molecular weight noncollagenous protein rich in aspartic acid, glutamic acid and leucine, named Ch-21 [5], but the protein has not been identified in humans.

Bone Markers

Most of the organic matrix of bone (90%) is constituted of collagen I. The remaining 10% is accounted for by minor quantities of other collagen types (III, V, VI and XII) and by several noncollagenous components, the main ones being the two gamma-carboxylated proteins, bone and matrix Gla protein. Other minor constituents include cell attachment proteins, such as fibronectin, thrombospondin, osteopontin and bone sialoprotein I (osteonectin), proteoglycans and growth factors [3]. As shown in table 1, thus far only some of these constituents have been exploited as turnover markers. Bone formation markers include the carboxy-terminal propeptide of collagen I, bone Gla protein (osteocalcin); while bone resorption markers include hydroxyproline, hydroxylysine glycosides and pyridinium crosslinks. In addition, enzymes required for osteoblast (alkaline phosphatase bone isoenzyme) and osteoclast

Table 1. Main biochemical markers of cartilage and bone turnover

	Formation	Resorption
Cartilage	collagen II C-propeptide	chondroitin sulfate
Bone	collagen I C-propeptide	hydroxyproline
	collagen I N-propeptide	hydroxylysine glycosides
	osteocalcin	pyridinium crosslinks
		Type I collagen crosslinked telopeptides
	bone specific alkaline phosphatase	tartrate-resistant acid phosphatase

activity (tartrate-resistant acid phosphatase) have also been used as turnover markers. The rest of the paper will discuss the application of these markers to growth studies.

Bone Formation Markers

Osteocalcin, regarded as the main formation marker, is synthesised in osteoblasts as a propeptide. After cleavage of the signal peptide, it is carboxylated, in the presence of vitamin K. This allows the protein to bind calcium and to assume the conformation needed to bind the hydroxyapatite lattice in bone. About one third of the secreted osteocalcin remains free and appears in the circulation, from where it is rapidly cleared by the kidney. Serum also contains osteocalcin fragments originated by bone resorption, noncarboxylated osteocalcin and – according to some authors – the propeptide. The clinical use of serum osteocalcin is supported by its positive correlation with bone formation histomorphometric parameters in osteoporosis and in patients undergoing longterm parenteral nutrition. Serum osteocalcin measurements have been performed in several studies on metabolic bone disease [6].

Evidence of a correlation with growth rates comes from the observation that serum osteocalcin levels parallel the height velocity curve, with higher values in childhood and during adolescence, that later fall to adult values [7]. When growth is arrested, such as in PEM, osteocalcin levels fall: in 6 Senegalese children aged 11–24 months affected by kwashiorkor, baseline osteocalcin was only 12% of healthy controls; after 1 week of nutritional rehabilitation it increased significantly, and after 3 weeks it was no longer significantly different from controls [8]. In other conditions of retarded growth, such as growth hormone deficiency [9] and idiopathic short stature [10], serum osteocalcin is also reduced, compared to age-matched controls. Treatment with growth hormone restored normal values, both in the pre-pubertal and the pubertal [11]. The authors claim that the increment in osteocalcin observed after 3 months of treatment contributed to the prediction of growth response after 12 months of treatment, but the extra variance explained was actually less than 1%. However, no correlation between serum osteocalcin and growth rates was observed by Michaelsen et al. [12], who studied 91 infants from birth to 12 months. In these children, both crown-heel and knee-heel length were measured at 1, 3, 9 and 12 months and the velocities correlated to the serum osteocalcin levels measured at 2, 6 and 9 months. Exclusively breast-fed infants had much higher osteocalcin levels at 2 months, while their growth rate between 1 and 3 months was the same as that of the other groups.

Although they are useful, osteocalcin measurements should be carefully interpreted. First of all, osteocalcin has a diurnal rhythm, with an early morning peak, so that standardisation of the blood sampling time is required; the diurnal pattern may be altered by hormonal treatment, and perhaps the daily integrated profile should be studied. Second, circulating osteocalcin is in equilibrium with hydroxyapatite-bound osteocalcin, and the equilibrium is affected by the vitamin K status of the individual and by drug- or disease-induced alterations in the affinity for bone [6]. Third, the clearance of the protein is affected by the rate of glomerular filtration that may be impaired in elderly subjects, and by its renal catabolism that may also be defective [6]. The fourth and perhaps most important caveat in the use of serum osteocalcin values is that of the assay specificity. The measure of osteocalcin is in fact affected by technical problems: most commercial antibodies recognise also fragments deriving from bone resorption, thus leading to wide scatter of values in data obtained with different assays [13]. Samples should be handled with great care, as osteocalcin is sensitive to freeze-thawing cycles, to haemolysis and to the presence of calcium in the biological material.

The propeptide of osteocalcin might be a more specific bone formation marker. Recently, Kanzaki et al. [14] have been able to detect the osteocalcin propeptide in serum, contradicting previous observations that the propeptide was entirely metabolised intracellularly [15]. As the propeptide can only come from osteoblasts its measurement in biological fluids would be specifically related to bone formation rates. In growth-homone-deficient children serum pro-osteocalcin levels measured after 1 month of treatment with growth hormone were correlated to height velocity during the following 12 months.

The carboxy-terminal propeptide of collagen I (PICP) is another widely used formation marker; in fact, the excision of procollagen propeptides occurs extracellularly and their serum level carefully reflects collagen I synthesis. The use of PICP as a bone marker is based on the assumption that, although collagen I is also present in skin, the mass and turnover rate of bone are such that most procollagen I should come from skeletal tissues. However, this assumption has never been verified experimentally. The age profile of serum PICP is parallel to the curve of height velocity. Saggese et al. [16] were able to point out a pubertal peak at 10–12 years in girls and 14–16 years in boys. Trivedi et al. [17] showed that serum levels of PICP could predict the response to growth hormone therapy. In prepubertal children with growth disorders treated with growth hormone, the percent increase in serum procollagen observed after 3 months was correlated to the increase in height velocity at 12 months [17]. Serum PICP was lower than controls in children affected by growth retardation as a consequence of long-term corticosteroid treatment; after treatment with growth hormone, PICP changes anticipated changes in growth velocity [18].

The N-terminal propeptide of type III collagen (PIIINP), an indicator of fibrous tissue metabolism, has also been used as a growth marker. This application is supported by the finding that serum levels parallel the growth velocity curve, also showing a pubertal peak [17]. Serum levels of PIIINP do not discriminate between growth-hormone-deficient children, short normals, normals and tall children [17], but they are able to anticipate the 6-month growth response after the first 5 weeks of growth hormone therapy [19]. However, changes in PIIINP levels are unlikely to result from changes in bone turnover, since collagen III is present in bone in minimal quantities.

Alkaline phosphatase is a zinc-containing enzyme needed to hydrolyze pyrophosphate and to activate the mineralisation process and has been used as a bone marker for a long time. Several isoforms, characterised by different carbohydrate side chains, are present in the body: the bone isoenzyme (BAP) constitutes 60–70% of the total alkaline phosphatase in serum and can be separated by HPLC or lectin precipitation. BAP has an age-related profile, similar to other formation markers [20], but its measure is relatively new and has not yet been applied to growth research.

Bone Resorption Markers

Hydroxyproline excretion has been among the first collagen metabolites to be used as a marker of bone resorption, but has been progressively abandoned due to its lack of specificity. Plasma hydroxyproline derives from many sources other than extracellular collagen, and only about 10% of plasma hydroxyproline is excreted in the urine, the rest being metabolised by the liver;

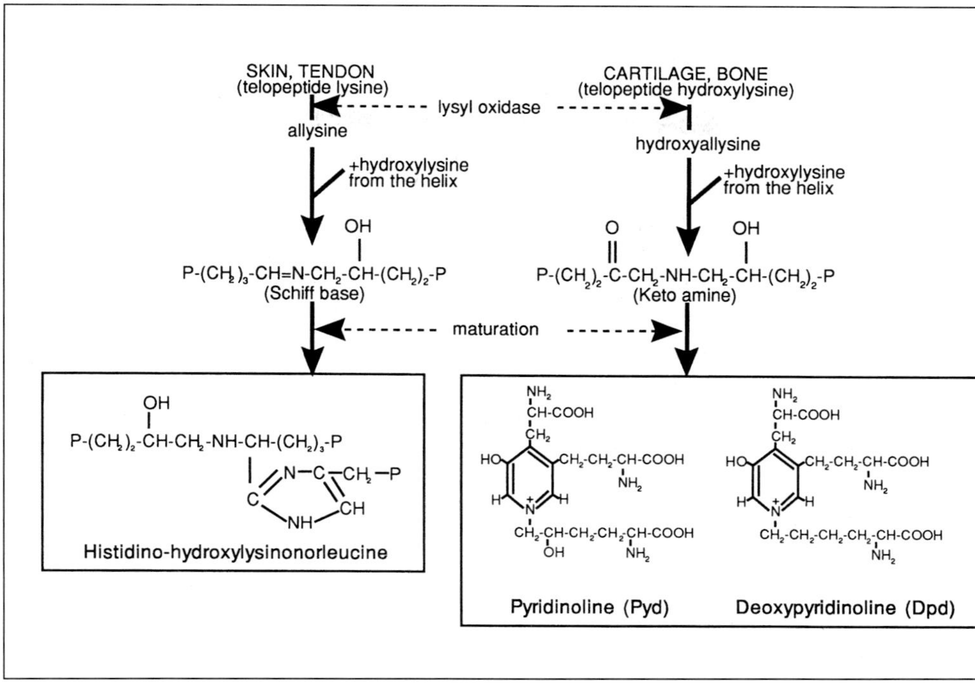

Fig. 1. Biochemical pathway of collagen crosslink synthesis. Modified from Robins [1].

only at high levels of bone turnover does urinary hydroxyproline correlate with more specific markers of bone collagen breakdown [21]. Other metabolites, such as the hydroxylysine glycosides, are instead more specific degradation products, since they only originate from collagen and are not degraded in the body. The monosaccharide, galactosyl-hydroxylysine is mostly present in bone and has been used to indicate bone resorption in metabolic bone disease, but not yet in growth research [22].

A specific indication of resorption rates of bone collagen comes from the study of pyridinium crosslinks, the covalent intermolecular bonds of mature collagen [23]. The biochemical pathway leading to crosslink formation is shown in figure 1. Once collagen fibrils have assembled in the matrix led by electrostatic interactions, they have to be stabilised by covalent crosslinking. The first step involves the oxidative deamination of ε-amino groups of lysine or hydroxylysine residues located in the telopeptide portions of the molecule; lysyl oxidase converts them to reactive aldehydes. Lysine, that is more common in skin, originates in allysine-based crosslinks, while hydroxylysine yields hydroxyallysine-based compounds, typical of cartilage and bone. These aldehyde derivatives crosslink with lysine or hydroxylysine from the helix and form initially labile crosslinks and then insoluble keto-amines. During the so-called maturation process, two or three collagen molecules join and form tri-functional crosslinks characterised by a 3-hydroxypyridinium ring. The tri-functional crosslink involving hydroxylysine is pyridinoline (Pyd), and the one involving lysine is deoxypyridinoline (Dpd). Mature collagen crosslinks have a relatively high tissue specificity. Pyd is present in cartilage, ligaments, tendons and bone collagen, but is absent from skin. Dpd appears in bone, dentin, aorta and ligaments, and is virtually absent from cartilage [24]. Bone is a high turnover tissue, particularly in children, and has a large mass, compared to other tissue also containing crosslinks; therefore, bone should be the primary source of crosslinks. Collagen crosslinks are quantitatively excreted in urine, where they appear as free and peptide-bound. The only source of urinary crosslinks is endogenous collagen metabolism, because collagen crosslinks present in the diet are not absorbed [25].

Collagen crosslinks can be measured in serum and urine. In the HPLC method developed by Black et al. [26] crosslinks are first extracted by partition chromatography, either from an acid hydrolysate (allowing quantification of both free and peptide-bound forms) or straight from urine. The eluate is then analyzed by reverse-phase HPLC, allowing the separation of Pyd and Dpd. This method has been later improved by the addition of an internal standard and by its automation [27]. Pyridinium crosslinks can also be measured by immunochemistry [28], and assays have also been developed against the carboxyterminal [29] and aminoterminal [30] telopeptides of collagen I, both containing crosslinks.

Urinary Pyd and Dpd have gained consensus as markers of metabolic bone disease and the evaluation of treatment [31]. At the Human Nutrition Unit of the National Institute of Nutrition, we have carried out several studies to evaluate their application to growth research, in collaboration with Dr. Simon Robins, of the Rowett Research Institute. Methodological problems, crucial in the validation of growth markers, have been tackled first. Not unlike the adults [32], crosslink output in children shows a nyctohemeral rhythm, with an early morning peak [33]; timed collections are thus required. Crosslink output is also variable between days, so that up to three collections are required to monitor the excretion over a week [33]. In children, standardisation for creatinine may not be suitable, as muscle mass and possibly muscle turnover are prone to rapid changes. A study performed on 41 children aged 3–5, in whom urine collections and anthropometry was performed once every month for 5 consecutive months, showed that mass of the skeleton, predicted with a formula including height and bone diameters, explained about 50% of the variance in crosslink excretion [33]. As the cubic power of height also showed the same relationship, in successive studies crosslink output was expressed as nmol/h/height (m) cubed. Using spot urinary collections and expressing the results as a ratio to urinary creatinine, Shaw et al. [34] could not show a relationship with growth velocity.

We then studied urinary levels of crosslinks in children of different ages. The results are shown in figure 2 and refer to a group of 108 children aged 3–5 and to a second group of 164 children aged 6–18 years, all of them with normal growth. For reasons of space, only Pyd values are shown, but Dpd output follows the same profile. Pyd and Dpd were higher in the 3 to 5-year-old children, decreased slowly between 8 and 10–12 years, increased slightly at puberty (14 years in girls and 16 years in boys) and later dropped to adult values. The correlation with age has also been shown by Rauch et al. [35].

We then moved to analyze the relationship between crosslink output and height velocity, and we carried out a 5-month longitudinal follow-up of the 3 to 5-year-old children. Crosslink output was significantly correlated to height velocity measured over the 30 days following the urinary collections, although only a small proportion of the variance was explained (fig. 3a). The correlation with height velocity in longer observation periods was higher ($r=0.78$), but it could be demonstrated only when repeated measures of crosslink excretion were averaged over the whole period of observation (fig. 3b). Changes in height velocity between the first and the second trimester of observation were also correlated to changes in urinary crosslink output [36].

The relationship with height velocity also appeared in studies on abnormal growth. In collaboration with Prof. Michael Golden, we studied 58 malnourished boys (age 13 ± 6 months; weight-for-height: -2.4 ± 0.8 Z-score; height-for-age: -3.6 ± 1.6 Z-score) at admission and after nutritional rehabilitation [37]. The rate of height gain during hospitalisation was related to the children's crosslink excretion at admission. Crosslink output, standardised by height cubed, could explain on its own about 14% of the variance in height gain.

We also studied 7 children affected by growth hormone deficiency, 13 children affected by idiopathic short stature, and 7 girls affected by Turner's syndrome. At baseline, the children were all prepubertal, had height-for-age below –2.5 Z score and growth velocity below the 25th centile for bone age, measured over the past year. The children were then treated for 12 months with 15 IU/m^2/week human recombinant growth hormone. Crosslink output at diagnosis was not significantly different from that of healthy controls, and

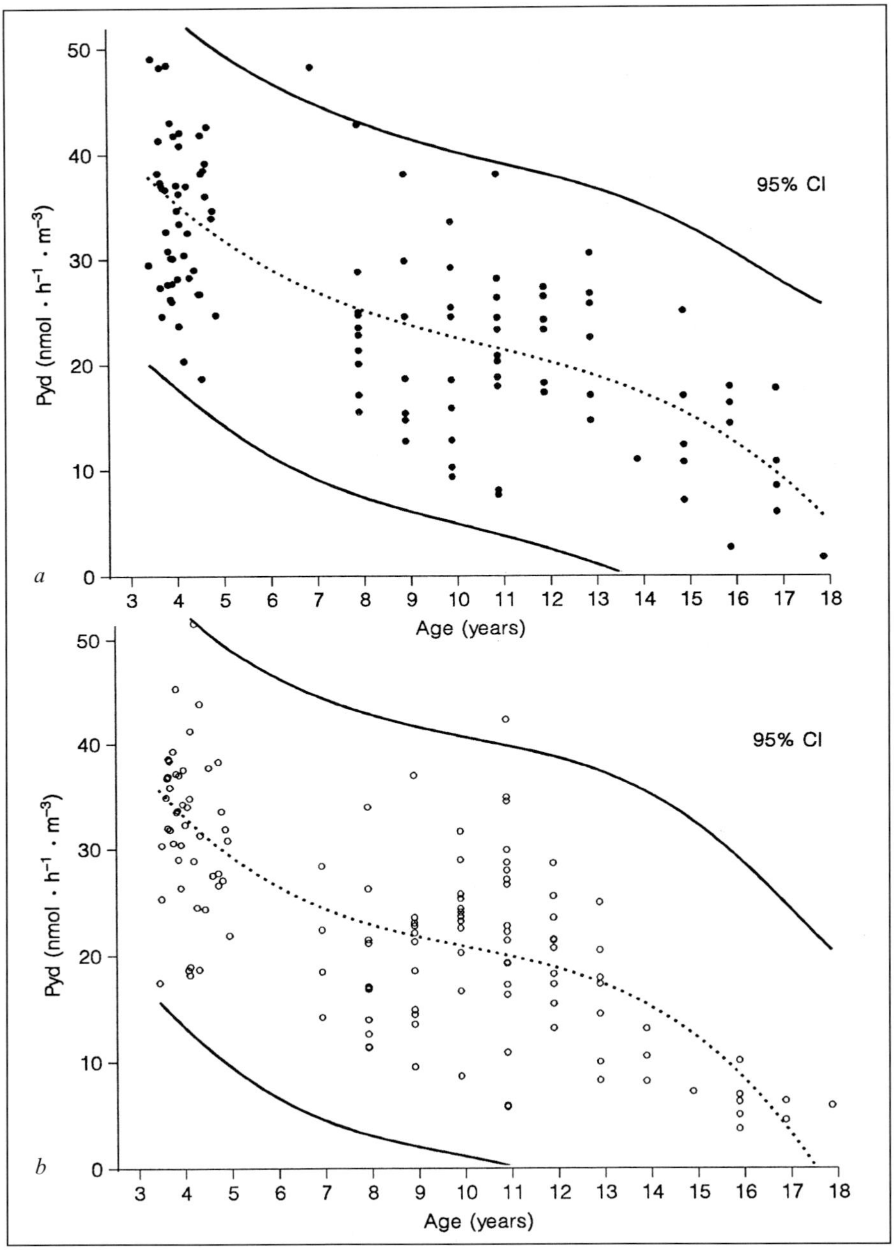

Fig. 2. Age changes of crosslink excretion. A third degree polynomial (dotted line) with the 95% Confidence Interval (solid lines) calculated from the individual values is also shown. *a* Boys. *b* Girls.

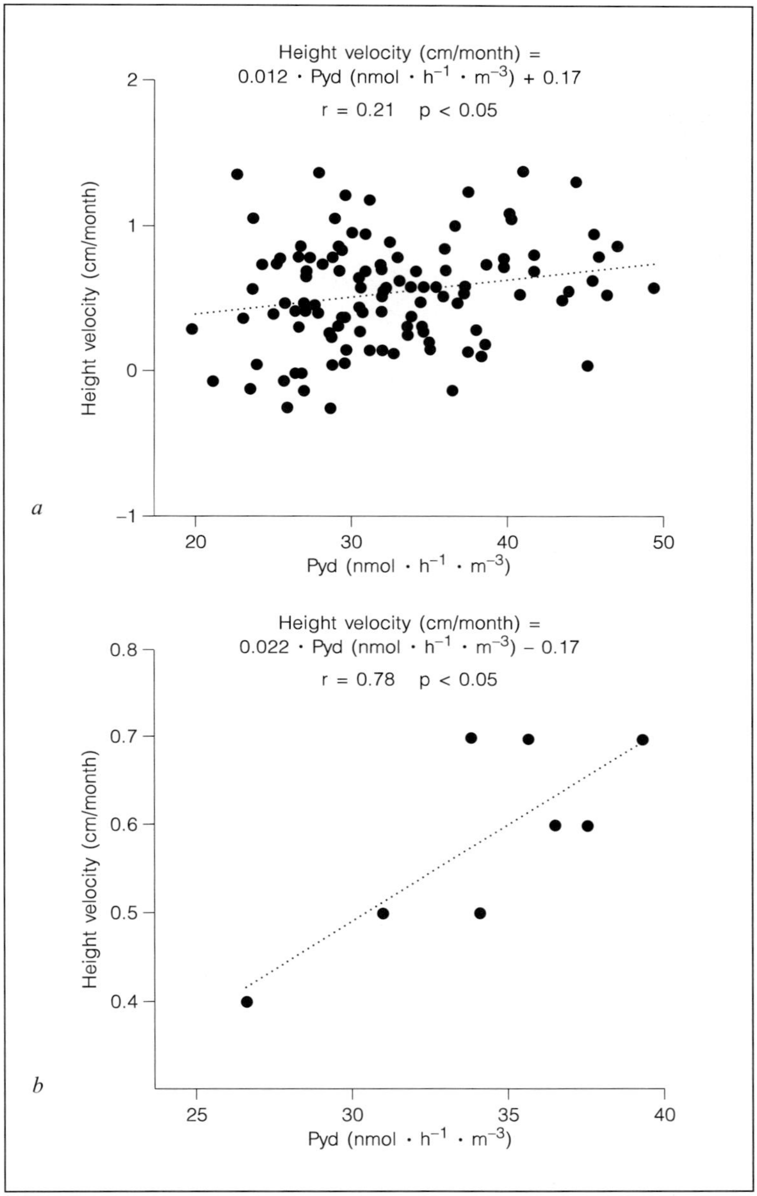

Fig. 3. Relationship between crosslink excretion and height velocity in healthy children. *a* 1-month height velocity (each point is the mean of 3 observations). *b* 5-month height velocity (each point is the mean of 15 observations).

it increased with growth hormone therapy. In growth hormone deficiency and in idiopathic short stature, the change in excretion after the first month could be used to predict growth performance, measured as the change in height velocity from pretreatment values, explaining about 50% of the variance [38]. However, the correlation was not present in Turner's syndrome [33].

These studies showed that collagen crosslinks are useful to indicate bone turnover rates in children. However, we need to recognise at least two limitations: the first one, common to all resorption markers, is that they measure only the remodelling component of skeletal growth. Usually, lengthening and remodelling are geared to each other, but there may be conditions in which increased resorption is not parallelled by increased formation and increased lengthening, such as Turner's syndrome. The second limitation is related to the kinetic of crosslink formation, about which there is very limited information. Future research should verify whether the rate of formation of reducible crosslinks and their conversion to mature forms is influenced by growth rates. Indeed, we know that collagen turnover is inversely proportional to the solubility, so that young collagen is degraded more than mature one. This would prevent the identification of the early phases of catch-up growth, or would lead to an underestimate of the growth potential.

Finally, an enzyme synthesised by osteoclasts (tartrate-resistant acid phosphatase) has also been used to evaluate bone resorption [39], but its presence in macrophages has questioned its specificity.

Conclusion

Awaiting for the development of a specific marker measuring the activity of growth cartilage, existing markers can be used, provided their limitations are accounted for. Among formation markers, a research protocol should include osteocalcin, measured with double-antibody assays, pro-osteocalcin – if the preliminary results are confirmed – and perhaps the carboxyterminal propeptide of collagen I. Among resorption markers, Pyd and Dpd appear the best choice. The limitations of single markers can be partially overcome by the combined use of formation and resorption markers, that may be usefully associated to accurate auxological anthropometry, such as the measure of the knee-heel length.

Markers of bone metabolism may provide an important contribution to nutrition research, not only for their contribution to elucidate the mechanisms leading to stunted growth, but also for their possible application to the study of nutrition requirements of normal and accelerated growth.

References

1 Robins SP: Biochemical markers for assessing skeletal growth. Eur J Clin Nutr 1993;47(suppl 1): 199–209.

2 Kember NF: The physiology of the growth plate; in Hernandez M, Argente J (eds): Human Growth: Basic and Clinical Aspects. Amsterdam, Elsevier, 1992, pp 81–86.

3 Heinegard D, Oldberg A: Structure and biology of cartilage and bone matrix noncollagenous macromolecules. FASEB J 1989;3:2042–2051.

4 Kirsch T, Von Der Mark K: Isolation of human type X collagen and immunolocalization in fetal human cartilage. Eur J Biochem 1991;196:575–580.

5 Descalzi Cancedda F, Manduca P, Tacchetti C, Fossa P, Quarto R, Cancedda R: Developmentally regulated synthesis of a low molecular weight protein (CH 21) by differentiating chondrocytes. J Cell Biol 1988;107:2455–2463.

6 Hauschka PV, Lian JB, Cole DE, Gundberg CM: Osteocalcin and matrix Gla protein: Vitamin K-dependent proteins in bone. Physiol Rev 1989;69:990–1047.

7 Gundberg CM, Lian JB, Gallop PM: Measurements of τ-carboxyglutamate and circulating osteocalcin in normal children and adults. Clin Chim Acta 1983;128:1–8.

8 Proudhon C, Sall G, Ndiaye B, Lemonnier D: Régulation nutritionnelle de l'ostéocalcine sérique: Étude dans le kwashiorkor. C R Acad Sci Paris III 1991;313:233–238.

9 Johansen JS, Giwercman A, Hartwell D, Nielsen CT, Price PA, Christiansen C, Skakkebaek NE: Serum bone Gla-protein as a marker of bone growth in children and adolescents: Correlation with age, height, serum insulin-like growth factor I, and serum testosterone. J Clin Endocrinol Metab 1988;67:273–278.

10 Colle M, Ruffie A, Ruedas E: Osteocalcin in children with short stature. Acta Paediatr Scand 1988; 333(suppl):196–197.

11 Johansen JS, Jensen SB, Riis BJ, Rasmussen L, Zachmann M, Christiansen C: Serum bone Gla protein: A potential marker of growth hormone (GH) deficiency and the response to GH therapy. J Clin Endocrinol Metab 1990;71:122–126.

12 Michaelsen KF, Johansen JS, Samuelson G, Price PA, Christiansen C: Serum bone gamma-carboxyglutamic acid protein in a longitudinal study of infants: Lower values in formula-fed infants. Pediatr Res 1992;31:401–405.

13 Delmas PD, Christiansen C, Mann KG, Price PA: Bone Gla protein (osteocalcin) assay standardization report. J Bone Miner Res 1990;1:5–11.

14 Kanzaki S, Hosoda K, Moriwake T, et al: Serum propeptide and intact molecular osteocalcin in normal children and children with growth hormone (GH) deficiency: A potential marker of bone growth and response to GH therapy. J Clin Endocrinol Metab 1992;75:1104–1109.

15 Gundberg CM, Clough ME: The osteocalcin propeptide is not secreted in vivo or in vitro. J Bone Miner Res 1992;7:73–80.

16 Saggese G, Bertelloni S, Baroncelli GI, Di Nero G: Serum levels of carboxyterminal propeptide of type I procollagen in healthy children from 1st year of life to adulthood and in metabolic bone disease. Eur J Pediatr 1992;151:764–768.

17 Trivedi P, Risteli J, Risteli L, Hindmarsh PC, Brook CGD, Mowat AP: Serum concentrations of the type I and III procollagen propeptides as biochemical markers of growth velocity in healthy infants and children and in children with growth disorders. Pediatr Res 1991;30:276–280.

18 Allen DB, Goldberg BD: Stimulation of collagen synthesis and linear growth by growth hormone in glucocorticoid-treated children. Pediatrics 1992;89:416–421.

19 Tapanainen P, Risteli L, Knip M, Käär M-L, Risteli J: Serum aminoterminal propeptide of type III procollagen: A potential predictor of response to growth hormone therapy. J Clin Endocrinol Metab 1988;67:1244–1249.

20 Kuwana T, Sugita O, Yakata M: Reference limits of bone and liver alkaline phosphatase in the serum of healthy subjects according to age and sex as determined by wheat germ lectin affinity electrophoresis. Clin Chim Acta 1988;173:273–280.

21 Robins SP: Turnover and crosslinking of collagen; inWeiss JB, Jayson MIV (eds): Collagen in Health and Disease. Edinburgh, Churchill-Livingstone, 1982, pp 160–178.

22 Moro L, Modricky C, Rovis L, De Bernard B: Determination of galactosyl hydroxylysine in urine as a means for the identification of osteoporotic women. Bone Miner 1988;3:271–276.
23 Robins SP: Crosslinking of collagen: Isolation, structural characterization and glycosylation of pyridinoline. Biochem J 1983;215:167–173.
24 Robins SP, Duncan A, Riggs BL: Direct measurement of free hydroxy-pyridinium crosslinks of collagen in urine as new markers of bone resorption in osteoporosis; in Overgaard K, Christiansen C (eds): Osteoporosis. Copenhagen, Osteopress Aps, 1990, pp 465–468.
25 Colwell A, Russell RGG, Eastell R: Factors affecting the assay of urinary 3-hydroxypyridinium crosslinks of collagen as markers of bone resorption. Eur J Clin Invest 1993;23:341–349.
26 Black D, Duncan A, Robins SP: Quantitative analysis of the pyridinium crosslinks of collagen in urine using ion-paired reversed-phase high-performance liquid chromatography. Anal Biochem 1988; 169:197–203.
27 Pratt DA, Daniloff Y, Duncan A, Robins SP: Automated analysis of the pyridinium crosslinks of collagen in tissue and urine using solid-phase extraction and reversed-phase high-performance liquid chromatography. Anal Biochem 1992;207:168–175.
28 Robins SP: An enzyme-linked immunoassay for the collagen crosslink, pyridinoline. Biochem J 1982;207:617–620.
29 Risteli J, Elomaa I, Niemi S, Novamo A, Risteli L: Radioimmunoassay for the pyridinoline cross-linked carboxy-terminal peptide of type I collagen : A new serum marker of bone collagen degradation. Clin Chem 1993;39:635–640.
30 Hanson DA, Weis MAE, Bollen A-M, Masian SL, Singer FR, Eyre DR: A specific immunoassay for monitoring human bone resorption: Quantitation of type I collagen cross-linked telopeptides in urine. J Bone Miner Res 1992;7:1251–1258.
31 Robins SP, Black D, Paterson CR, Reid DM, Duncan A, Seibel MJ: Evaluation of urinary hydroxypyridinium crosslink measurements as resorption markers in metabolic bone diseases. Eur J Clin Invest 1991;21:310–315.
32 Mc Laren A, Isdale AH, Whiting PH, Bird HA, Robins SP: Physiological variations in the urinary excretion of pyridinium crosslinks of collagen. Br J Rheumatol 1993;32:307–312.
33 Branca F: Biochemical markers of skeletal growth in children; PhD thesis, University of Aberdeen, 1995.
34 Shaw NJ, Dutton J, Fraser WD, Smith CS: Urinary pyridinoline and deoxypyridinoline excretion in children (abstract). Bone and Tooth Society Winter Meeting, London, 1993.
35 Rauch F, Schonau E, Woitge H, Remer T, Seibel M: Urinary excretion of hydroxypyridinium cross-links of collagen reflects skeletal growth velocity in normal children. Exp Clin Endocrinol 1994; 102:94–97.
36 Branca F, Golden MH, Robins SP, Ferro-Luzzi A: Urinary output of collagen cross-links as a growth marker in children. Bibl Nutr Dieta, Basel, Karger, 1994, vol 51, pp 113–115.
37 Branca F, Robins SP, Ferro-Luzzi A, Golden MHN: Bone turnover in malnourished children. Lancet 1992;340:1493–1496.
38 Spagnoli A, Branca F, Spadoni GL, et al: The effect of growth hormone on serum and urinary markers of bone and connective tissue in children with severe growth retardation (abstract). International Symposium on Growth and Growth Disorders, Florence, 1993.
39 Piedra C, Torres R, Rapado A, et al: Serum tartrate-resistant acid phosphatase and bone mineral content in postmenopausal osteoporosis. Calcif Tissue Int 1989;45:58–60.

Francesco Branca, Istituto Nazionale della Nutrizione,
via Ardeatina, 546, I-00179 Rome (Italy)

Porrini M, Walter P (eds): Nutrition in Pregnancy and Growth.
Bibl Nutr Dieta. Basel, Karger, 1996, No 53, pp 103–108

Body Composition of Italian Children

Roberto Bellù, Maria T. Ortisi, Enrica Riva, Marcello Giovannini

Department of Paediatrics, University of Milan, Italy

Growth in humans is very complex: not only are bone, muscle and fat growing, but also, and more importantly, heart, liver and brain.

It is now becoming evident that factors affecting growth may produce effects in very complicated systems and functions peculiar to the human species.

Investigating infant body composition is a difficult task. Despite many methods being available for adults, only few are feasible in newborns and infants [1]. Dilution (doubly labelled water and others) and plicometry are the most used methods, but new techniques have been developed in the last few years. In this paper we present the results and comment on the data obtained with one of these new techniques, namely TOBEC (Total Body Electrical Conductivity). This new method measures the perturbance of a fixed, weak electromagnetic field by the human body [2–4]. The signal is proportional to the fat-free mass that, hence, can be measured [5–7].

One important point of TOBEC measurement is that despite there being many approaches for validating new methods for measuring body composition in adults, the choice is very limited when dealing with infants and newborns because of the lack of a 'gold standard' [1].

Three major approaches can be considered: (1) comparison with reference data; (2) development of animal model(s); (3) comparison with other techniques.

All these points have been carried out extensively by Fiorotto [6] in Houston. Although much work is still to be done to validate the TOBEC technique, the correlation coefficients are good (table 1) and the unexplained points are very low when comparing TOBEC measurements with other techniques [7–9]. Considering our own data of body composition measurement in the first year of life, a good agreement between the reference data of Fomon et al. [10] and our estimates was evident. On this basis, we could develop our reference chart for body fat and fat-free mass in infants, relying on the measurement of children aged 0–6 months (table 2).

Table 1. Correlation of TOBEC-derived estimates of fat-free mass

	Total body K	Densitometry	TBW $H_2^{18}O$
Piglet	0.99		
Infant			0.94
Adolescent	0.97	0.99	
Adult	0.94	0.99	

Table 2. Body composition and anthropometric parameters of children from birth to 6 months

	Weight g	FFM g	Fat g	Fat %	Height cm
Birth					
Preterm, AGA	2,411	2,107	304	12.65	47.30
Term, SGA	2,272	2,020	252	10.93	47.50
Term, AGA	3,187	2,724	463	14.34	50.40
Term, LGA	4,159	3,414	745	17.48	54.00
Four months					
Preterm, AGA	5,522	4,126	1,396	24.80	58.80
Term, SGA	5,750	4,100	1,649	28.30	58.50
Term, AGA	6,537	4,699	1,835	28.00	62.40
Term, LGA	7,966	5,422	2,544	31.50	66.30
Six months					
Preterm, AGA	7,438	5,353	2,085	28.03	66.00
Term, SGA	7,020	5,002	2,018	28.75	65.00
Term, AGA	7,451	5,184	2,267	30.43	66.00
Term, LGA	10,000	6,675	3,325	33.25	72.00

AGA = Appropriate for gestational age; SGA = small for gestational age; LGA = large for gestational age; FFM = fat-free mass.

Given the encouraging results obtained with the TOBEC techique, we performed a series of studies in different clinical and epidemiological settings.

(1) We investigated the body composition of newborns during their first 6 months of life, splitting the sample into four groups: (a) preterm (33–37 weeks gestational age), with appropriate birth weight (AGA); (b) full-term, small for gestational age (SGA); (c) full-term, healthy; (d) full-term, large-for-gestational

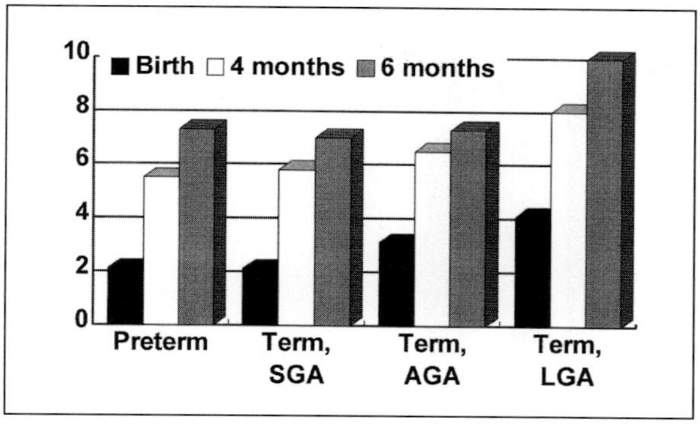

Fig. 1. Body weight (kg) in selected newborns.

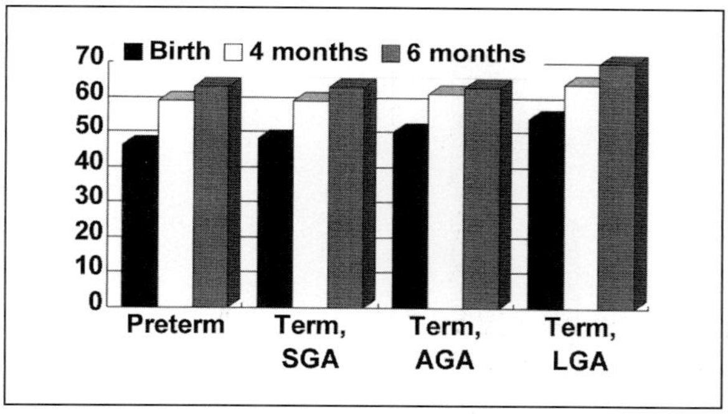

Fig. 2. Length (cm) in selected newborns.

age (LGA). The data for length, weight, fat-free mass and fat mass are shown in figures 1–4 and table 2. It becomes evident that the major differences in body weight and body composition disappear at 6 months; LGA babies show higher body weight and fat mass at 6 months.

Similar data have been shown by Venkataraman using the DEXA technique.

(2) In the clinical setting, we studied the body composition of 10 PKU children, diagnosed at birth and treated early with low fenylalanine diet.

As compared to normal children, patients with phenylketonuria show lower body fat during the first 6 months of life; they show catch-up growth and normal body fat by the age of 12 months. This pattern of fat deposition reflects the initial problem of 'tuning' the optimal diet for each child and the good nutritional value of the diet during the first year of life.

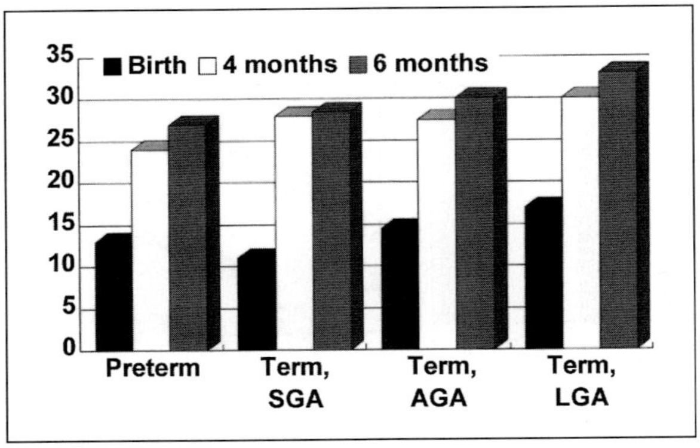

Fig. 3. Body composition (fat%) in selected newborns.

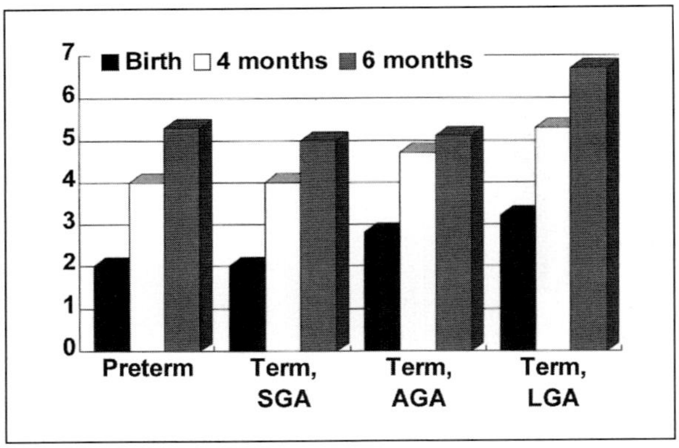

Fig. 4. Fat-free mass (kg) in selected newborns.

(3) A further issue is the use of body measurement for evaluating the effect of different types of feeding in the normal infant population. Data from the DARLING Study pointed out that breast-fed babies are leaner than formula-fed babies when estimating body fat according to skinfold measurements. Eighty-one 12-month-old infants, exclusively breast-fed for at least 6 months and bottle-fed or breast-fed for less than 2 months were enrolled into the study. ANOVA was used to compare the variables. Weight, stature and cranial circumference were found to be lower among breast-fed infants (table 3). Both groups did not differ significantly in terms of fat-free mass, but body-fat mass was comparatively lower among breast-fed infants. Multiple regression analysis showed that a significant correlation exists between the

Table 3. Sample characteristics and anthropometric indices versus type of feeding (average ± SD)

Variable	Formula-fed (n = 53)	Breast-fed ≥6 months (n = 26)	p
Birthweight, g	3,312 ± 477	3,127 ± 427	0.01
Week of gestation	39.7 ± 1.1	39.3 ± 1.2	
Introduction of solid foods, months	5.0 ± 0.8	4.8 ± 0.8	
Introduction of cow's milk, months	7.77 ± 3.0	9.5 ± 2.8	0.05
Length, cm	77.22 ± 2.38	75.45 ± 3.26	0.01
Cranial circumference, cm	46.67 ± 1.50	45.90 ± 1.28	0.04
Weight, kg	10.04 ± 1.06	9.37 ± 0.92	0.01
Weight gain over 12 months, kg	6.73 ± 0.95	6.25 ± 0.84	0.05
FFM, kg	7.36 ± 0.64	7.02 ± 0.66	
Fat, kg	2.68 ± 0.55	2.35 ± 0.46	0.02
Fat, %	26.51 ± 3.31	25.00 ± 3.36	0.05
BMI, kg/m^2	16.82 ± 1.41	16.46 ± 1.07	

Table 4. Correlation between selected variables and fat mass (%)

Variable	R	p
Skinfold		
Triceps	0.857	<0.001
Bicipital	0.666	<0.001
Subscapular	0.751	<0.001
Iliac	0.661	<0.001
Arm fat area	0.841	<0.001
BMI	0.738	<0.001
Weight	0.622	<0.001

Fat (%) = 2.135 + B1 * Tric.
B1 = 2.842, ES(B1) = 0.373.

percentage of body fat and the duration of breast-feeding and that a lower percentage of body fat correlated with longer breast-feeding.

The significant difference in body mass between breast-fed and formula-fed infants largely shows as an accretion of body-fat mass. This is relevant in particular for the energy and nutrient requirements of formula-fed infants who have a definite increase in, but a slightly higher metabolically active body mass. Thus, adaptive metabolic responses to breast-feeding and formula-feeding can be investigated in terms of different body compositions according to dietary regimen. A long-term aim of a study from TOBEC-derived measurements is to assess whether fat mass accretion remains high among formula-fed infants over the following years. A TOBEC-type methodology, equally applicable in successive paediatric stages, can contribute to our understanding of infant growth patterns in relation to the early mode of feeding.

(4) The last point concerns the correlation between TOBEC-derived measurement of fat mass and plicometry.

In a study performed on 40 healthy children (age range: 0–8 months), we measured four skinfolds and body fat with TOBEC. The correlation coefficients between each skinfold and TOBEC-derived fat mass are reported in table 4. The coefficients look quite good, thus indicating that the two methods measure the same thing; nevertheless, too much variance is unexplained.

In conclusion, new methods for estimating body composition in infants are now available [1]. Although there is a need for further definition of their validity and reliability, these new techniques are likely to help better define the nutritional status and needs of younger children, both from the epidemiological point of view and in the clinical setting.

References

1 Klish WJ: Methods for measuring body composition in infants and children: The 'Gold Standard'; in Klish WJ, Kretchmer N (eds): Body Composition Measurements in Infants and Children. 98th Ross Conf Pediatric Research. Columbus, Ross Laboratoires, 1989.

2 Van Itallie TB, Segal KR, Funk RC: Total body electric conductivity: A rapid measure index of lean body mass: Human body composition and fat distribution; in Norgan NG (ed): Euro-Nut Report No 8. Wageningen, Euro-Nut, 1986.

3 Van Loan M, Mayclin P: A new TOBEC instrument and procedure for the assessment of body composition: Use of Fourier coefficients to predict lean body mass and total body water. Am J Clin Nutr 1987;45:131–137.

4 Fiorotto ML, Cochran WJ, Klish WJ: Fat-free mass and total body water of infants estimated from total body electrical conductivity measurements. Pediatr Res 1987;22:417–421.

5 Fiorotto ML, Cochran WJ, Funk RC, Sheng HP, Klish WJ: Total body electrical conductivity measurements: Effects of body composition and geometry. Am J Physiol 1987;252:R794–R800.

6 Fiorotto ML: The application of TOBEC (Total Body Electrical Conductivity): Measurements for estimating the fat and fat-free mass of human infants; in Klish WJ, Kretchmer N (eds): Body Composition Measurements in Infants and Children. 98th Ross Conf Pediatric Research. Columbus, Ross Laboratoires, 1989.

7 Cochran WJ, Klish WJ, Wong WW, Klein PD: Total body electrical conductivity used to determine body composition in infants. Pediatr Res 1986;20:561–564.

8 Forbes GB: Methods for determining composition of human body. Pediatrics 1962;29:477.

9 Brans YW: Estimation of total body water by $H_2{}^{18}O$ dilution in neonates. Clin Res 1988;36:896.

10 Fomon SJ, Haschke F, Ziegler EE, Nelson SE: Body composition of reference children from birth to age 10 years. Am J Clin Nutr 1982;35:1169–1175.

Prof. Marcello Giovannini, Scuola di Spez. in Pediatria III, Clinicà Pediatrica V,
Università' degli Studi di Milano, Via A. di Rudini, 8, 20142 Milano (Italy)

Porrini M, Walter P (eds): Nutrition in Pregnancy and Growth.
Bibl Nutr Dieta. Basel, Karger, 1996, No 53, pp 109–115

Relationship of Nutritional Status and Functional Development in Preschool Age Children

Jana Pařízková

Laboratory of Health Promotion, 4th Department of Internal Medicine,
1st Medical School, Charles University, Prague, Czech Republic

Introduction

There are relatively few integrated comprehensive studies of preschool children. Anthropometric, or nutritional and/or functional aspects of development were usually studied separately, and this also applies to motor and/or psychological development. Early age is considered one of the critical periods when positive intervention may result in more favorable development later in life. But also the reverse is possible, i.e. deficient and/or imbalanced nutrition of varying character, lack of motor or psychological stimulation, hypokinesia, etc., may cause deterioration not only of the current, but also of the future development of functional capacity and health. All that also applies to the above-mentioned natural factors such as dietary intake and physical activity.

Repeated measurements of preschool children were performed in late Czechoslovakia since 1972; both cross-sectional and longitudinal measurements of morphological, functional, nutritional, biochemical and psychological variables were made in a total of twelve studies [1–6]. Selected results mainly from the last study in 1989 will be presented.

Subjects and Methods

9,587 boys and girls (A) aged 4–6 years from Bohemia and Moravia (now Czech Republic) were examined. A questionnaire concerning the social, environmental, family, health, somatic and motor developmental aspects of children was completed by both parents

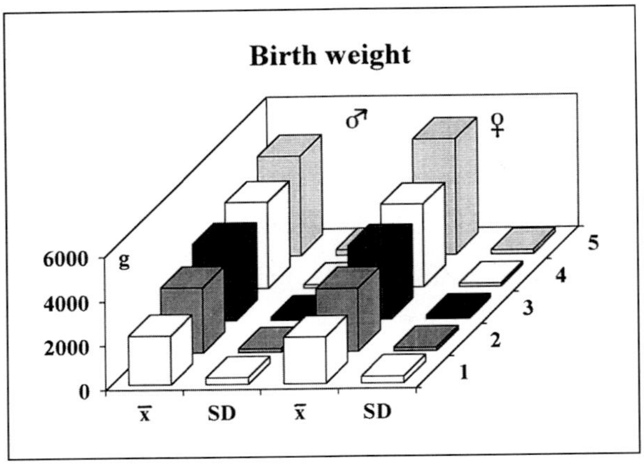

Fig. 1. Average values of birth weight of 1,005 children (B) subdivided into five categories.

and kindergarten teachers. A subsample of 1,005 children (B) was examined in greater detail as regards somatic development (height, weight, body mass index, Z-score for height, weight, proportionality, circumferential measurements, body posture) and motor performance (20 m dash, 500 m run-and-walk, broad jump, ball throwing, skill tests) [2, 5–7]. The research was implemented with the help of kindergarten teachers and district physical education instructors (i.e. graduates of Pedagogic and Physical Education Faculties of Charles University) who participated in a special course in our laboratory. Dietary intake was measured using questionnaires (7 days) completed by parents and kindergarten teachers, and analyzed using a computer programme prepared for our country in smaller samples of children (n = 87). The results statistically analyzed by Ing. Z. Roth. Statistical Department of the State Health Institute in Prague ($\bar{X}$, SD, SE, $V_{\bar{X}}$, analysis of variance, hierarchical log-linear analysis, correlations, etc.).

Results and Discussion

Mean values of height, weight, body mass index, circumferential measurements and motor performance did not differ significantly from the standard values assessed earlier [1, 3, 5, 7] and/or from the standards of WHO and other countries.

Therefore, we focussed our attention first on the impact of environmental factors such as living in the area of high and/or low pollution, health factors (i.e. risk pregnancy, health problems of the infant resulting in the necessity of regular medical check-ups until the end of the first year of life), beginning of independent walking, etc. However, at the age of 4–6 years the assessed vari-

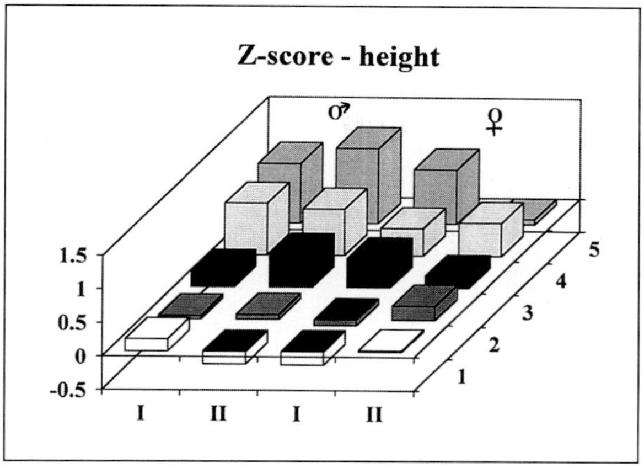

Fig. 2. Z-score for height in preschool boys and girls, under (I) and above 5 years of age (II), subdivided according to their birth weight into five categories. Analysis of variance – significance of factors: birth weight.

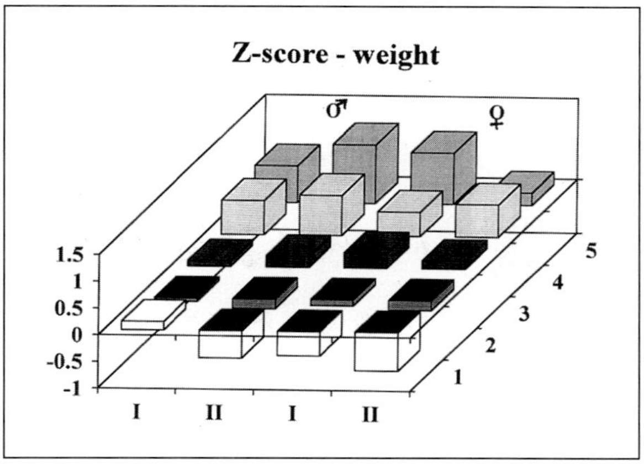

Fig. 3. Z-score for weight of preschool boys and girls in age and birth weight subgroups. Analysis of variance – significance of factors: birth weight.

ables did not differ significantly regarding the factors mentioned above in groups A and B.

The analysis of variance and hierarchical log-linear analysis showed a significant impact of birth weight only, differentiated into five subgroups (fig. 1; B). In addition, children were always subdivided into two age subgroups, i.e. under and above 5 years (I and II). Average values of weight, height, Z-scores

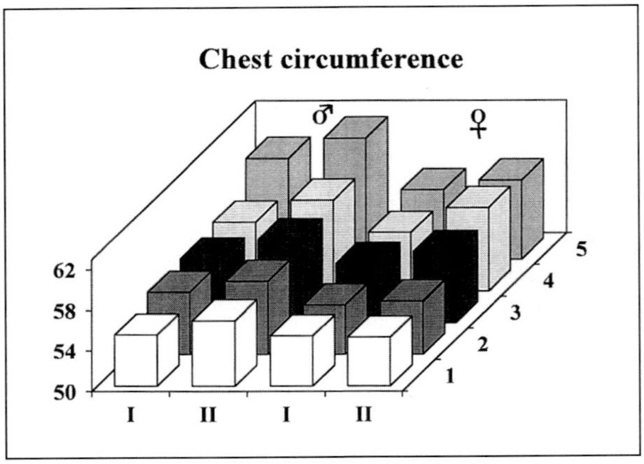

Fig. 4. Average values of chest circumference in preschool boys and girls, in age and birth weight subgroups. Analysis of variance – significance of factors: birth weight, age, sex.

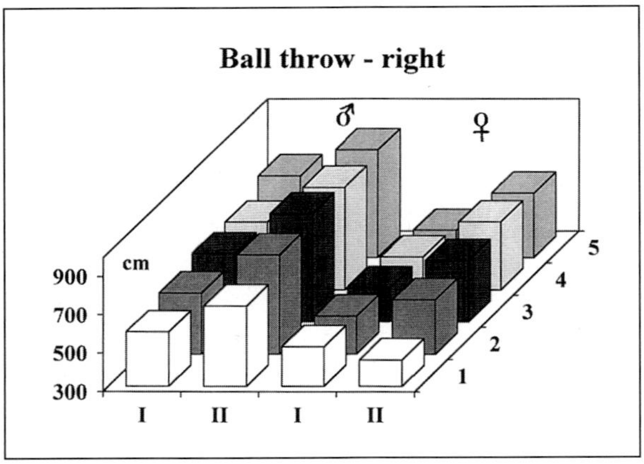

Fig. 5. Average values of ball throwing by the right hand in preschool boys and girls, in age and birth weight subgroups. Analysis of variance – significance·of factors: age, sex.

for height and weight, and circumferential measurements (chest, waist, hips) varied significantly according to birth weight, sex and age categories (fig. 2–4). But body mass index did not differ according to birth weight.

As regards motor performance, the analysis of variance showed significant differences only in left-hand ball throwing. The throw performance by the right hand, eventually influenced by previous exercise and training, tended to

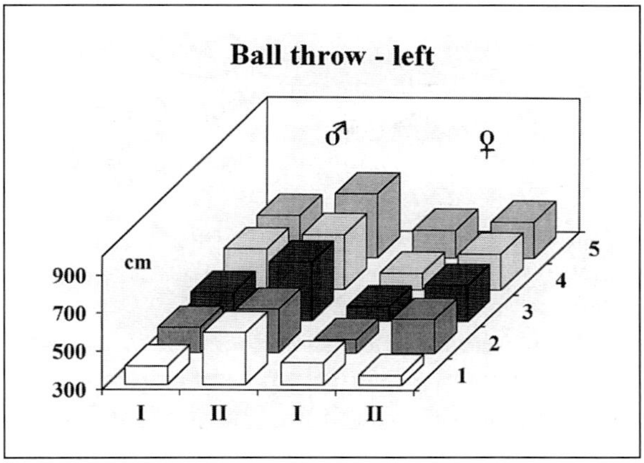

Fig. 6. Average values of ball throwing by the left hand in preschool boys and girls, in age and birth weight categories. Analysis of variance – significance of factors: birth weight, age, sex.

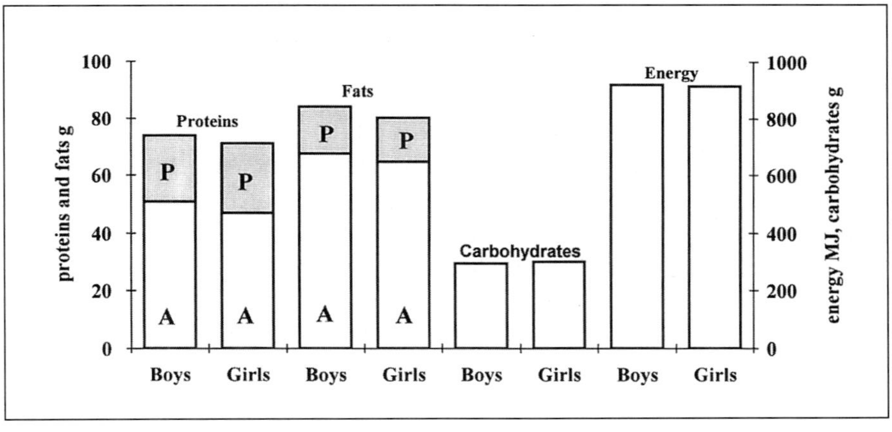

Fig. 7. Daily intake of energy, proteins, fats (total: A = animal; P = plant) and carbohydrates in 87 boys and girls 4–6 years of age (energy MJ/100).

vary similarly, but the impact of birth weight was not significant (fig. 5, 6). The results in the long jump, and then in the 20 m dash and 500 m run-and-walk were not related to birth weight and actual body weight, and only occasionally a trend towards better results appeared in children born and remaining heavier.

When analyzing the groups of children in total, those born with higher birth weight started to walk sooner, had lower morbidity, and showed better results in some sensorimotor tests [8, 9]. Enrollment in regular physical education was less frequent in children born and remaining smaller in body size. Balance test (standing on one leg for 10 s) was significantly better in girls born heavier (B).

Food intake measured in smaller subsamples was generally higher than recommended dietary allowances (RDA), both international and national. This was mostly due to a higher intake of fats (fig. 7).

Larger body size in itself was a predisposition for a higher level of functional capacity provided the lifestyle was favorable, including an increased level of physical activity. Previous assessments of serum lipids showed significantly higher HDL levels along with the trend for larger body size, lower fatness, better results in modified step test, motor and sensorimotor tests and higher food intake in children with a higher level of spontaneous physical activity [10]. Other measurements showed also better performance along with larger body size and better body posture in children enrolled in regular physical education specially tailored and organized for this age period, i.e. exercises for preschool children together with one of the parents, and/or grandparents [2, 4, 6, 7].

As follows from the results mentioned above, the functional capacity and fitness including motor skills, endurance, cardiorespiratory efficiency, body size, serum lipids, etc., may be positively influenced very early in life. All factors which reduce the prevalence of low birth weight and increase it to an optimal level are considered positive as regards later development of the child. But larger body size itself especially when accompanied by increased fat deposition, which correlates significantly with a higher serum level of cholesterol ($r = 0.448$, $0.01 < p < 0.02$) and triacylglycerols ($r = 0.494$, $0.01 < p < 0.02$) at preschool age already [10], may be considered more of a handicap, especially when bearing in mind the delayed health consequences later in life [11, 12]. Therefore, positive interventions in lifestyle of the whole family concerning diet and physical activity regimes from the beginning of life seem to be of utmost importance.

References

1 Pařízková J, Čermák J, Horná J: The impact of ecological factors and physical acivity on somatic and motor development of preschool children; in Shepard RJ, Lavallée H (eds): Physical Fitness Assessment: Principles, Practice and Application. Springfield. Thomas, 1978, pp 238–247.

2 Pařízková J: Body Fat and Physical Fitness. The Hague, Martinus Nijhoff BV/Medical Division, 1977.

3 Pařízková J, Čermák J, Horná J: Sex differences in somatic and functional characteristics of preschool children. Hum Biol 1977;49:437–451.

4 Pařízková J: Longitudinal study of somatic and functional development of preschool children; in Somogyi JC, deWijn JF (eds): Nutritional Aspects of Physical Performance. Bibl Nutr Dieta, Basel, Karger, 1979, vol 27, pp 65–76.
5 Pařízková J, Adamec A: Longitudinal study of anthropometric, skinfold, work and motor characteristics of boys and girls three to six years of age. Am J Phys Anthrop 1980;50:387–396.
6 Pařízková J: Growth, functional capacity and physical fitness in normal and malnourished children; in Bourne G (ed): International Nutrition in Health and Disease. World Rev Nutr Diet 1987;51: 1–44.
7 Pařízková J, Adamec A, Berdychová J, Čermák J, Horná J, Teplý Z: Growth, Fitness and Nutrition in Preschool Children, Prague, Charles University, 1983.
8 Pařízková J: The impact of environment on somatic and motor development of preschool children; in Claessens A, Lefevre J, Vanden Eynde B (eds): World-Wide Variation in Physical Fitness, Leuven, Institute of Physical Education. Katholieke Universiteit, 1993, pp 131–135.
9 Pařízková J: Nutrition and physical performance during growth and development. Proc XVth Int Congr Nutrition: IUNS, Adelaide; in Wahlquist ML, et al (eds): Nutrition in a Sustainable Environment. London, Smith-Gordon-Nishimura, 1994, pp 463–465.
10 Pařízková J, Macková E, Kábele J, Macková J, Škopková M: Body composition, food intake, cardiorespiratory fitness, blood lipids and psychological development in highly active and inactive preschool children. Hum Biol 1986;58:261–273.
11 Prevention in Childhood and Youth of Adult Cardiovascular Diseases. Report of WHO Expert Committee. Tech Rep Ser 792, Geneva, World Health Organization, 1990.
12 Pařízková J: Nutrition, Physical Activity, and Health in Early Life. Boca Raton, CRC Press, 1996.

Prof. Jana Pařízková, MD, PhD, Sc, 4th Department of Internal Medicine,
Laboratory Health Promotion, U Nemocnice 2, 128 06 Prague 2 (Czech Republic)

Porrini M, Walter P (eds): Nutrition in Pregnancy and Growth.
Bibl Nutr Dieta. Basel, Karger, 1996, No 53, pp 116–128

Fat Nutrition during Infancy and Childhood

Kurt Widhalm

University of Vienna, Department of Pediatrics, Vienna, Austria

Introduction

Although the velocity of research on lactation, human milk and infant nutrition has slowed, work on milk lipids continued because milk fatty acids can be altered by dietary as well genetic factors. Furthermore, the interest on fat intake in infant nutrition has raised considerably in the last few years since prevention of later cardiovascular diseases, which are the main causes of deaths in industrialized countries, is already being taken into consideration in infancy and childhood. Finally, the story of polyunsaturated fatty acids, in particular from the long-chain type and their possible effects on the development of the central nervous system, especially in the premature infant, has focused the interest on fatty acid consumption and their biological effects not only in premature infants but also in older children and adolescents.

Infant Nutrition by Human Milk

In order to ascertain the quantities of nutrients conveyed to the infant in breast milk, the amount of milk must be determined. Estimation of 24-hour milk intake is desirable, but often unattainable. There are two methods described by which the amount of milk intake can be measured: the deuterium oxide method and the doubly labelled water method [1].

With the latter method, it has been made possible to study the energy content of human milk without sampling the product: the measured water output is equivalent to water (milk or formula) intake.

In the study of Lucas et al. [2] 29 infants aged 3 months consumed 892 ml of formula per day and had a metabolizable energy content of 66 kcal/dl. The calculated energy was 68 kcal/dl. At 6 weeks of age, the metabolizable energy content for formula-fed infants was 60 kcal/dl. The amounts for breast-fed infants were 53 kcal/dl at 6 weeks and 58 kcal/dl at 3 months. These data had lower values than the accepted 70 kcal/dl [3].

Factors Affecting Total Lipid Content of Milk

The amount of lipids in human milk increases from 4.0% at 12 weeks to 5.5% at 16 weeks; thus, the energy content increases from 68.5 to 83.0 kcal/dl [4].

Few controlled investigations are available on changes that occur in the fat content of milk as a result of alterations in the diet. In general, the amounts of fat, protein and lactose are little changed by variations in diet. Allen et al. [5], reported that undernutrition can reduce fat content by 25%, and Hachey et al. [6], noted that the fat content of milk from 5 women was 2.5% on a low-fat diet (5% fat, 80% carbohydrates) as compared to 3.3% on a high-fat diet (40% fat, 45% carbohydrates). Other authors, however, were not able to observe any relationship between milk fat content and maternal fat intake either in terms of total fat intake or percent of calorie intake. They did not note a significant relation between milk fat content and maternal protein content in the later stages of lactation. The effects of several diseases on milk lipids have been examined.

The milk from one diabetic patient contained 2.95% fat: control milks contained 4.53% at days 6–7 postpartum [7]. However, the fat content in milk from diabetics later in lactation did not differ much from that of the controls.

The effect of the mother's amount of adiposity has a positive influence on the level of fat in milk. Nommsen et al. [8], obtained a positive and significant correlation between milk lipids concentration and maternal ideal body weight at 6, 9 and 12 months postpartum.

Composition of Fat in the Milk

The fat in milk is derived from fat in the blood or by synthesis in the mammary gland. The sources of fat in the blood are the diet via the gut, the liver and adipose tissue.

Dietary fatty acids are transported to the mammary gland in chyclomicrons from the intestine and VLDL from the liver while adipose tissue fatty acids travel bound to albumin. In the mammary gland, fatty acid synthesis is limited to 10:0–14:0 using glucose as the source of carbon. Diet and adipose tissue are sources of 16:0 and longer fatty acids.

Table 1. General composition (%) of human and bovine milk and some infant formulas

Component	Milk		Formulas				
	human (36 days)	bovine	1	2	3	4	5
Protein	1.0	3.4	1.52	1.5	1.5	1.6	1.5
Casein, % potein	40	82	40	82	40	0	40
Fat	3.9	3.3	3.8	3.6	3.6	3.3	3.4
Lactose	6.8	4.8	7.0	7.2	7.2	7.2	7.6
kcal/dl	72	75	68	68	68	65	67

Adapted from reference [11] and personal correspondence. 1 = Enfamil, Mead Johnson; 2 = Similac, Ross; 3 = SMA, Wyeth; 4 = good start, Carnation, reconstituted, protein is hydrolyzed whey, contains lactose and maltodextrin; 5 = NAN, Nestle. All are for healthy term infants and the proteins in all except 4 are bovine milk whey and casein.

Peak amounts of dietary fatty acids appear in milk about 8–10 h after consumption with an exponential decay from 48 to 72 h. About 10% of the total acids, primarily 10:0 to 14:0 are synthesized in the mammary gland, 60% are derived from tissue synthesis and adipose tissue, and 30% come from the diet. In one study the percentages of kcal in the diet were 17% for protein, 56% for carbohydrates and 27% for fat [9].

Hachey et al. [10] reported that consumption of a low-fat diet produced milk with a fat content of 2.2%, while consumption of a high-fat diet produced milk with a fat content of 3.28%. The amounts of fat produced per day per breast, however, were almost the same: 8.0 and 7.4 g. Milk fat content was inversely correlated with body water flux, i.e. volume adjusted for fat content.

Lipid Classes

Information on the general composition of human and bovine milk and several infant formulas is given in table 1.

The lipids in human milk are mostly triglycerides (98%), cholesterol (0.4%), phospholipids (1.3%) and traces of other lipids. The major triglycerides were composed of 16:0, 18:1 and 18:2 regardless of variations in dietary fat.

Human milk contains approximately 185 various fatty acids. Milk fat globules must be first altered by gastric lipase before the core TGs can be digested by pancreatic colipase in the small intestine. Predigestion by gastric lipase compensates for the relatively low concentrations of pancreatic colipase and bile salts in infants as compared to children and adults. Most of the activity designated as lingual is of gastric origin in the human.

Requirements for Polyunsatuated Fatty Acids

The requirements of humans for 18:2 ω–6, the original essential fatty acid, are well established, the minimum amount for infants is set at 3% (240 mg/dl) by the American Academy of Pediatrics [12].

Deficiency of EFA was established by abnormal triene/tetraene (20:30 ω–9/20: 4 ω–6) ratios in plasma lipids or erythrocyte phospholipids.

An abnormal ratio is greater than 0.4.

Uauy [13] recommended 500–700 mg of 18:2 ω–6 per kg of infant weight per day, with a maximum of 12%.

Nutrition and Growth/Body Composition

In the past two decades there has been considerable debate whether breast-feeding has any influence on the fat distribution and might be protective against obesity. Early studies indicated that bottle-fed infants tended to be fatter than breast-fed infants, but the formulas in use at that time were quite different from those used today. Results of later studies were mixed: although some continued to show that breast-fed infants were leaner than formula-fed infants, most did not.

Longer-term studies of children who had been breast-fed or formula-fed in early life generally showed no difference in fatness at 3–8 years of age. However, in most of these studies the mean duration of breast-feeding was relatively short.

The Davis Area Research on Lactation, Infant Nutrition and Growth (DARLING) was designed to collect data from matched cohorts of infants who were either breast-fed or formula-fed until >12 months of age and who were not given solid food until 4 months or later.

From this study it can be concluded that there is a significant difference in fatness between breast-fed and formula-fed infants with the greatest differences evident between 9 and 15 months of age (figure 1) [14].

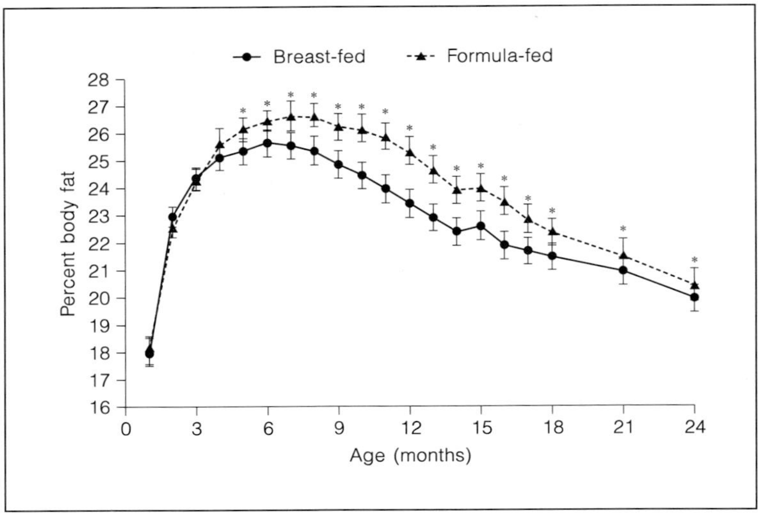

Fig. 1. Percent body fat ($\bar{x} \pm SE$) of breast-fed (n=41–46) and formula-fed infants (n=35–41). *p <0.05.

Long-Chain Polyunsaturated Fatty Acids (LCPUFA)

It is generally accepted that protein is relevant to body growth and minerals to skeletal growth. However, the main structural material of the brain is lipid, which has a unique profile of long-chain polyunsaturated fatty acids, which are essential fatty acids.

The brain has two unique characteristics, first it is 60% lipid and second, it has a remarkable high-energy consumption. During its most active phase in fetal growth, it uses as much as 70% of the energy delivered to it from the mother and postnatally it still uses 60% of the energy delivered. The fetus does not use fat as a source of LCPUFA postnatally as well as acting as an important energy buffer to help ensure that the needs of the developing brain are satisfied.

In this regard very fascinating studies have been published in the last 5 years, which should be reviewed briefly. The group of Hoffmann and Uauy were the first to recognize the structural role of long-chain polyunsaturated fatty acids and their functional correlates [15]. Thus, arachidonic acid as an important membrane component and precursor to prostaglandins, leucotrienes and epoxides have been described.

Besides the well-known deficiency symptoms for EFA subtle clinical symptoms including abnormal visual function and peripheral neuropathy have been

Table 2. Rod ERG results of preterm-infant study: 36 weeks postconception (x ± SD)

	Human milk (n = 10)	Corn oil (n = 19)	Soy oil (n = 24)	Soy + marine (n = 28)
log Threshold [1]	0.41 ± 0.59	1.08 ± 0.37*	0.71 ± 0.59	0.41 ± 0.61
log V_{max}, μV	1.20 ± 0.14	1.05 ± 0.11**	1.08 ± 0.20	1.22 ± 0.20
log k*	1.25 ± 0.54	1.73 ± 0.28**	1.39 ± 0.55	1.24 ± 0.47

[1] Scotopic troland-seconds.
Significantly different from human milk and soy + marine groups; * $p < 0.001$; ** $p < 0.005$.

reported. These complications may occur more in response to an insufficiency of the specific metabolic derivative of α-linolenic acid, namely docohexaenoic acid (DHA or cervonic acid 22:6:ω–3), which has been found in high concentrations in cerebral cortex and retina [16].

In a study to evaluate the necessity of dietary ω–3 LCPUFAs preterm infants were fed human milk, corn oil-based formula (ω–3 fatty acid deficient), soy-oil-based formula (rich in precursor fatty acids) or marine oil supplemented formula (containing docosohexaenoic acid).

It could be shown by these authors that there are significant differences in regards to the threshold for electroretinograms (table 2) and for the visual acuity (table 3) in the various dietary groups. The addition of marine oil improved eye functions in a similar manner to that observed in the breast-fed infants.

In another study by Carlson et al. [17] from the USA, it was also possible to show that preterm infants who were fed a marine oil-supplemented formula had an improved visual acuity by improving their DHA status (fig. 2).

Based on these experimental data and on the facts that the concentrations of cervonic and arachidonic acid is significantly lower in red blood cells in infants fed formula milks, it has been advocated to add both ω–6 and ω–3 LCPs to current artificial formulas [18].

Dietary Fat and Obesity in Childhood

The prevalence of obesity has increased dramatically during the past 15–20 years. From 1963 to 1980 obesity increased by 54% in the USA, from 18 to 27% among 6–11 year-olds and by 39%, from 16 to 22% among 12–17 year-olds [19].

Table 3. Visual-acuity results of preterm-infant study: 57 weeks postconception

	Human milk (n=9)	Corn oil (n=12)	Soy oil (n=16)	Soy +marine (n=13)
VEP				
logMAR[1]	0.51±0.27	0.67±0.15†	0.63±0.22*	0.40±0.25
Snellen[2]	20/65	20/95*	20/85	20/50
FPL				
logMAR	0.73±0.07	0.91±0.11†	0.85±0.13	0.79±0.10
Snellen	20/115	20/165	20/140	20/125

[1] $\bar{x} \pm SD$: log of the minimum angle of resolution.

[2] Mean Snellen equivalent, where 20/20 is the average for normally sighted adults and 20/70 for VEP and 20/115 for FPL are average for normally sighted full-term infants at 57 weeks postconception.

Significantly different from human-milk and soy + marine groups: $*p<0.05$.

Obesity has been regarded as a problem of nutrient imbalance. In view of a lack of association between energy intake and obesity, interest has been focused on the diet content. It has been shown that not only the total energy content of an obese person's diet contributes to their obesity, rather the dietary composition of fat, protein and carbohydrate.

In a recently published study Gazzaniga and Burns [20] were able to show in 53 preadolescent children (aged 9–11 years) that obese individuals always have a higher fat intake than nonobese persons (table 4).

On the other hand, there was a positive correlation for all fats for all children and the percentage of body fat and a negative correlation to the protein and carbohydrate intake (table 5).

In addition, a highly significant relationship between the percentage of body fat and the percentage of energy intake from dietary fat could be shown (fig. 3).

An inverse relationnship could be shown for the carbohydrate intake and body fat (fig. 4).

From these and other studies, it has been concluded that obese children do not consume more energy than nonobese ones, after adjusting for body weight. Interestingly, they are consuming a greater proporation of their overall energy in the form of total dietary fat and saturated, monounsaturated and

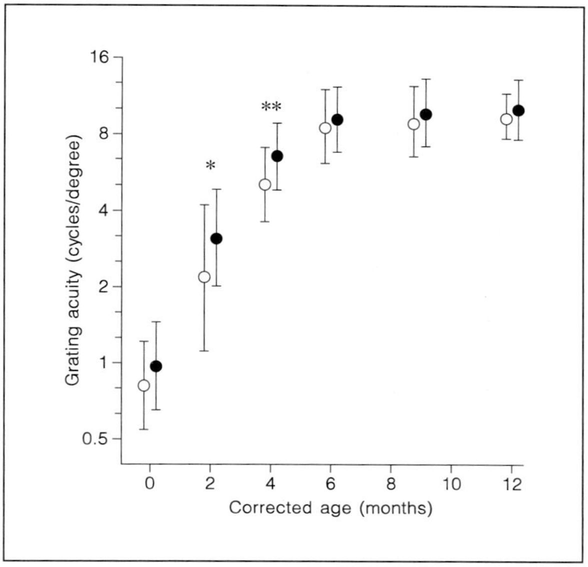

Fig. 2. Visual-acuity development in control (○) and marine oil-supplemented infants. Control infants differ from supplemented infants: * $p<0.014$, ** $p<0.002$. Individual acuities (cycles/degree) were converted to $\log_{10}$ cycles/degree and averaged for all infants in that formula age group, then plotted as $\bar{x} \pm SD$. The SDs in octaves (log SD/0.301) for control infants at corrected ages of 0, 2, 4, 6.5, 9, and 12 months were 0.58, 0.96, 0.48, 0.48, 0.38, and 0.30, respectively. The SDs for supplemented infants at corrected ages of 0, 2, 4, 6.5, 9, and 12 months were 0.58, 0.63, 0.47, 0.43, 0.45, and 0.39 octaves, respectively.

polyunsaturated fatty acids, and less in the form of carbohydrate than nonobese children.

Fat Intake and Hypercholesterolemia in Children

There is general agreement that elevated blood cholesterol acts as a major risk factor for coronary vascular diseases in adult populations, and the reduction of cholesterol levels decreases the risk. Because there are no comparable studies in childhood, the significance of cholesterol as a risk factor for coronary vascular disease must be inferred from less-indirect evidence.

Several distinguished panels like the American Academy of Pediatrics published recommendations that *all* children and adolescents should eat a diet which on average contains no more than 30% of total calories from fat, less than

Table 4. Nutrient intake, by sex and study group ($\bar{x} \pm SD$)

	Girls		Boys	
	Nonobese (n=16)	Obese (n=9)	Nonobese (n=14)	Obese (n=9)
Total fat, %[a]	34±6	39±6	31±6	39±5
Saturated fatty acid, %[b]	13±2	15±3	12±3	15±2
Monounsaturated fatty acid, %[c]	13±3	15±2	12±3	15±2
Polyunsaturated fatty acid, %	6±1	7±2	5±1	7±2
Sterol, phospholipid, %	2±0.5	3±0.4	2±1	3±0.4
Protein, %	13±2	13±1	15±3	13±2
Carbohydrate, %	53±7	49±5	55±8	49±6

Significant study-group effect (two-factor analysis of variance): [a] $p<0.001$; [b] $p<0.001$; [c] $p<0.025$.

Table 5. Correlation coefficients of percentage of body fat with nutrient intake

Percentage of body fat with	Combined groups (n=48)	Nonobese (n=30)	Obese (n=18)
Total fat, %	0.55[a]	0.41[b]	0.26
Saturated fatty acid, %	0.41[c]	0.23	0.12
Monounsaturated fatty acid, %	0.54[a]	0.45[c]	0.32
Polyunsaturated fatty acid, %	0.39[c]	0.37[b]	0.11
Protein, %	–0.12	0.08	0.04
Carbohydrate, %	–0.46[d]	–0.39[b]	–0.24

[a] $p<0.0001$; [b] $p<0.05$; [c] $p<0.01$; [d] $p<0.001$.

10% of total calories from saturated fat, and less than 300 mg of cholesterol per day [21].

There is no doubt that nutritional habits in most western countries have to be changed dramatically in order to achieve this goal. On the other hand, treatment of children having markedly elevated plasma cholesterol and LDL-cholesterol levels (as being affected with familial or other forms of genetic hypercholesterolemia) has to be carried out with a strict dietary regimen. In this regard, the substitution of animal protein by soy protein has been shown to lower LDL-cholesterol levels to a higher extent than a low fat/low saturated fat diet [22].

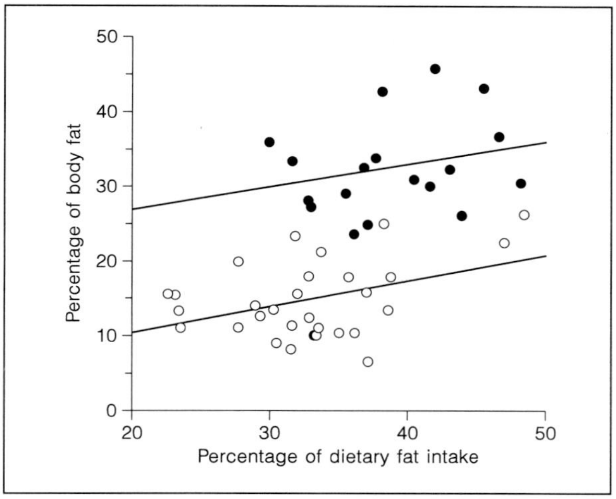

Fig. 3. Relationship between percentage of body fat and percentage of energy intake from dietary fat. ○ = Nonobese (r = 0.41, $p < 0.05$); ● = obese (r = 0.26, NS).

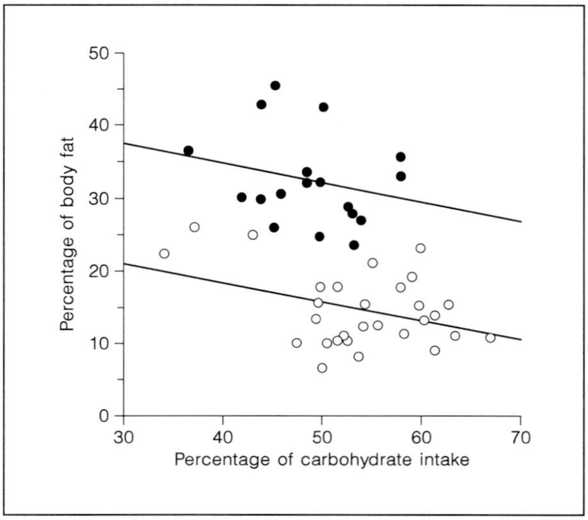

Fig. 4. Relationship between percentage of body fat and percentage of energy intake from carbohydrate. ○ = Nonobese (r = –0.39, $p < 0.05$); ● = obese (r = –0.24, NS).

Table 6. Proposals for food consumption in a balanced mixed diet [g (ml)/day]

	Children 4–6 years	Adolescents 13–14 years
Milk, milk products	350	450
Lean meat, sausages	50	80
Saltwater fish (1 × week)	100	200
Eggs (number/week)	2	2–3
Butter, margarine and vegetable oils	20	30
Bread (>50% wholegrain)	150	250
Cereals (wholegrain)	20	30
Potatoes	120	200
Vegetables	180	250
Fruit	180	250
Cake, biscuits, confectionery	≦40	≦80
Sugar, jam, preserves	≦25	≦45
Beverages	700	1,400
Energy content/day	1,500 kcal 6.3 MJ	2,500 kcal 10.5 MJ

On the other hand, there are observations showing that children who were placed on a very low fat diet due to hypercholesterolemia show marked growth and development problems [23].

In view of these reports it is very important to state that dietary counseling of children with different nutritional problems must be in the hand of well-trained, nutrition-experienced pediatricians.

Due to the fact that it is not very easy to meet all the requirements for all essential nutrients and to meet the recommendations for a low-fat diet for all children, researchers from the Institute for Nutrition Research in Dortmund tried to compile a nutrition program for children and adolescents (tables 6, 7) [24].

Conclusions

The quantity and quality of fat intake for the infant should be highly oriented on the composition of human milk. All infant formulas should be adapted as much as possible to the fat composition of human milk. There is a difference in regard to absorption between fats from human milk and adapted cows milk. For prematures and possibly for mature infants, the important

Table 7. Protein, fat and carbohydrates in the diets of children and adolescents (percent of energy)

	Present-day diet (1–14 years)	Proposed balanced mixed diet (≧2 years)
Protein, total, %	13	13
Animal:vegetable protein	66:33	50:50
Fat, total, %	39	35
Linoleic acid, %	4	7
Saturated fatty acids, %	18	13
P:S ratio	0.28	0.63
Cholesterol (mg/1,000 kcal)	170	100
Carbohydrates, total, %	48	52
Sucrose, %	14	8
Dietary fiber (g/1000 kcal)	8	16

P:S = Polyunsaturated:saturated fatty acids.

role of LCPUFA has been recognized. Therefore, infant formulas should be supplemented. Fat intake seems to influence the development of obesity. From the age of 2 years a prudent diet should not contain more than 30% fat. Low-fat diets can cause severe growth problems.

References

1 Fjeld CR, Brown KH, Schoeller DD: Validation of the deuterium oxide method for measuring average daily intake in infants. Am J Clin Nutr 1988;46:671–679.

2 Lucas A, Ewig G, Roberts SB, Coward WA: How much energy does the breast-fed infant consume and expend? Br Med J 1987;295:75–77.

3 National Research Council: Recommended Dietary Allowances, ed 10. Washington, National Academic Press, 1989, pp 1–385.

4 Ferris AM, Dotts MA, Clark RM, Ezrin M, Jensen RG: Macronutrients in human milk at 2, 12 and 16 weeks postpartum. J Am Diet Assoc 1988;88:694–697.

5 Allen JC, Keller RP, Neville MC, Archer P: Studies in human lactation: Milk composition and secretion rates of macronutrients in the first year of lactation. Am J Clin Nutr 1991;54:69–80.

6 Hachey DL, Silber GH, Wong WW, Garza C: Human Lactation. II. Endogenous fatty acid synthesis by the mammary gland. Pediatr Res 1989;25:63–68.

7 Bitman J, Hamosh M, Lutes V, Neville Mc, Seacat J, Wood DL: Milk composition and volume during the onset of lactation in a diabetic mother. Am J Clin Nutr 1989;50:1364–1369.

8 Nommsen LA, Lovelady CA, Heinig MJ, Lönnerdal B, Dewey KG: Determinants of energy, protein, lipid and lactose concentrations in human milk during the first 12 mo of lactation: The Darling study. Am J Clin Nutr 1991;53:457–465.

9 Hachey DL, Thomas MR, Enken EA, Garza C, Brown-Booth L, et al: Human lactation: Maternal transfer of dietary triglycerides labeled with stable isotopes. J Lipid Res 1987;28:1185–1192.

10 Hachey DL, Motil DJ, Wong WW, Garza C, Klein PD: Milk production and fat concentration are related to whole body water flux during human lactation. FASEB J 1989;3:A454.
11 Jensen RG, Ferris AM, Lammi-Keefe CJ: Lipids in human milk and infant formulas. Annu Rev Nutr 1992;12:417–441.
12 Committee on Nutrition, American Academy of Pediatrics: Fats and fatty acids; in Forbes GB, Woodruff CW (eds). Pediatric Nutrition Handbook. Elk Grove Village, American Academy of Pediatrics, 1985, vol II, pp 105–110.
13 Uauy R: Are ω–3 fatty acids required for normal eye and brain development in the human? J Pediatr Gastroenterol Nutr 1990;11:296–300.
14 Dewey KG, Heinig MJ, Nommsen LA, Peerson JM, Lönnerdal B: Breast-fed infants are leaner than formula-fed infants at 1 y of age: The DARLING study. Am J Clin Nutr 1993;57:140–145.
15 Uauy R, Treen M, Hoffmann DR: Essential fatty acid metabolism and requirements during development. Semin Pathol 1989;13:118–130.
16 Hoffmann DR, Birch EE, Uauy RD: Effects of supplementation with ω–3 long chain polyunsaturated fatty acids on retinal and cortical development in premature infants. Am J Clin Nutr 1993;85(suppl): 807S–812S.
17 Carlson SE, Werkman SH, Rhodes PG, Tolley EA: Visual acuity development in healthy preterm infants: Effect of marine oil supplementation. Am J Clin Nutr 1993;58:35–42.
18 Foreman van Drongelen MMHP, Westdorp JM, van Houwelingen AC, Hornstra G, Hasaart THM, Blanco CE: The essential fatty acid status of premature infants and the fatty acid composition of their diet. Am J Clin Nutr 1993;57(suppl):829S.
19 Gortmaker SL, Dietz WH, Sobol AM, Wehler CA: Increasing pediatric obesity in the United States. Am J Dis Child 1987;141:535–540.
20 Gazzaniga JM, Burns TL: Relationship between diet composition and body fatness, with adjustment for resting energy expenditure and physical activity in preadolescent children. Am J Clin Nutr 1993; 58:21–28.
21 American Academy of Pediatrics: Statement on cholesterol. Committee on Nutrition. Pediatrics 1992;90:469–473.
22 Widhalm K, Brazda G, Schneider B, Kohl S: Effect of soy protein diet versus standard low fat, low cholesterol diet on lipid and lipoprotein levels in children with familial or polygenic hypercholesterolemia. J Pediatr 1993;123:30–34.
23 Lifshitz F, Moses N: Growth failure: A complication of dietary treatment of hypercholesterolemia. Am J Dis Child 1989;143:537–542.
24 Kersting M, Schöch G: Achievable guidelines for food consumption to reach a balanced fat and nutrient intake in childhood and adolescence. J Am Coll Nutr 1992;11:74S–78S.

Prof. Dr. K. Widhalm, University of Vienna, Department of Pediatrics,
Währinger Gürtel 18–20, A–1090 Vienna (Austria)

Porrini M, Walter P (eds): Nutrition in Pregnancy and Growth.
Bibl Nutr Dieta. Basel, Karger, 1996, No 53, pp 129–134

Energy Cost of Growth in Infants

Eric Jéquier

Institute of Physiology, University of Lausanne, Switzerland

Introduction

Optimal growth of infants and children requires an adequate amount of energy and a balance of essential nutrients. Growth is characterized by an increase in the lean body mass and the fat mass, which consists of a deposition of protein and fat, respectively. Dietary carbohydrates are mainly used as an energy fuel and are oxidized after having been temporarily stored as glycogen in liver and muscles. Furthermore, the conversion of carbohydrate into fat is not an important pathway in humans [1]. Thus, the energy cost of growth represents the cost of deposition of protein and fat. The energy cost of growth (E_g) includes two components: the energy deposited in newly synthesized tissues or energy stored (E_s), and the energy expended to support the metabolic cost (E_m) of growth (fig. 1). The latter is the energy required to synthesize new molecules such as proteins, DNA, RNA, triglycerides and phospholipids.

Two approaches have been used to estimate the energy cost of growth of infants:

(1) The theoretical method, based on the change in body composition, which takes into account the energy equivalents of the nutrients deposited and the known energy cost for synthesizing the new molecules deposited.

(2) The clinical physiology approach which is based on measurements of energy balance in growing infants.

Energy Cost of Growth as Estimated from Body Composition Changes

The energy cost of growth is mainly dependent on the composition of weight gain. Because the composition of weight gain changes throughout the first year of life, the energy stored (E_s) is particularly dependent on the fat gain

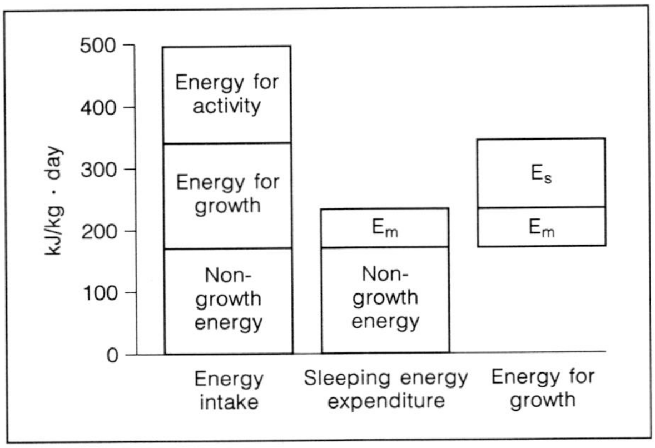

Fig. 1. Components of energy intake and energy expenditure. E_s = Energy stored; E_m = energy spent for the synthesis of new tissues or metabolic cost of growth.

(37.6 kJ/g fat). Body composition changes were assessed from measurements of total body water by deuterium dilution and of lean body mass by whole body counting of ^{40}K [2]. Table 1 shows that fat gain in boys and girls reaches a maximal value between the first and the third month of life, with an E_s of 18 kJ/g. Thereafter, the value of E_s declines to about 6 or 7 kJ/g at 1 year of age. This reflects the fact that the composition of weight gain is about 20% fat and 13% protein at 1 month, 40% fat and 10% protein at 3 months, and 10% fat and 17% protein at 12 months of life.

Using multiple-regression analysis, Roberts and Young [3] showed that to deposit 1 kJ of fat, E_m was 0.17 kJ, whereas to deposit 1 kJ of protein, E_m was 1.38 kJ. Thus, the energetic efficiency defined as the ratio of:

Energy deposited/total energy used for the depositon

is about 1/1.17 = 0.85 or 85%, for fat, and 1/2.38 = 0.42 or 42%, for protein.

The values of E_m for fat (0.17 kJ/kJ fat deposited) and for protein (1.38 kJ/kJ protein deposited) are explained by the ATP utilization for digestion, transport, processing of the nutrients, and the energy cost of synthesizing lipids and proteins, respectively [4]. For protein deposition as new lean tissues, E_m includes the high rate of protein turnover needed to promote net protein deposition in infants. In newborn infants and in premature infants, about 5 g of protein are synthesized for each gram deposited [5, 6]. The major processes that require energy in the form of high-energy phosphate bonds are not only those involved in peptide bond synthesis, but amino acid transport, RNA turnover, phosphorylation of enzymes, GTP-GDP exchange proteins (signal transduction), second messen-

Table 1. Energy cost of growth derived from body composition studies

	Age months	Weight gain g/day	Fat gain g/day	Protein gain g/day	E_g kJ/day	E_s kJ/g
Boys	0–1	29.3	6.0	3.7	288	10
	1–2	35.2	14.1	3.5	590	17
	2–3	29.9	12.9	3.0	536	18
	3–4	20.8	8.3	2.3	351	17
	4–5	16.6	5.5	2.0	240	14
	5–6	15.2	4.1	2.0	188	12
	6–9	12.6	1.8	2.0	101	8
	9–12	10.7	1.0	1.8	68	6
Girls	0–1	26.0	5.6	3.3	266	10
	1–2	28.6	12.8	2.8	529	18
	2–3	24.3	10.1	2.6	424	17
	3–4	18.6	7.3	2.1	310	17
	4–5	16.1	5.9	1.9	254	16
	5–6	15.0	4.9	1.9	216	14
	6–9	11.2	1.7	1.8	94	8
	9–12	10.0	1.2	1.7	74	7

Adapted from Fomon et al. [2].

gers (phosphatidyl inositol system) and even protein degradation are all processes which are energy dependent. In addition, not only the quantity of total protein formed and degraded is important, but also the qualitative nature of the proteins synthesized should be considered [7].

Energy Cost of Growth as Estimated by Clinical Physiological Studies

The total energy cost of weight gain has been determined from the relationship between metabolizable energy intake (MEI) and weight gain. The slope of the regression of MEI on weight gain represents the sum of the increments in both energy storage and in total energy expenditure expressed per gram weight gain. By using this approach and excluding figures published by Ashworth [8], values of E_g have ranged between 18 and 25 kJ/g of weight gain in infants from 0 to 18 months (table 2) [8–16]. This method, however, tends to overestimate the real cost of growth, since it includes the energy spent for physical activity (E_{pa}).

Table 2. Energy balance studies to compute the energy cost of growth

Reference	Age months	Weight kg	Weight gain g/kg·day	E_g kJ/g	E_s kJ/g	E_m kJ/g
Infants recovering from malnutrition						
Ashworth [8], 1969	16.5	5.1	10.0	40.2	–	–
Kerr et al. [9], 1973	12.9	–	–	25.1	19.7	–
Spady et al. [10], 1976	12.2	5.2	8.4	18.4	13.8	4.6
Jackson et al. [11], 1977	14.8	6.8	–	–	25.9	–
Infants from a developing country						
Stettler et al. [12], 1992	0.5	3.6	10.6	25.7	20.2	5.5
Preterm infants						
Brooke et al. [13], 1979	0–2	1.9	13.7	23.8	16.7	7.1
Reichman et al. [14], 1982	0–1	1.3	16.8	20.5	18.0	2.8
Gudinchet et al. [15], 1986	0–2	1.4	11.2	–	–	2.2
Freymond et al. [16], 1986	0–1	1.7	16.6	10.3	–	–

Adapted from Butte et al. [17].

A second approach corrects for E_{pa} and therefore is a more meaningful method to assess E_g. This method consists of measuring the energy stored (E_s) by computing the difference between MEI and total daily energy expenditure (TDEE). The value of E_s is then expressed per g of weight gain; experimental values of E_s between 13 and 20 kJ/g have been reported for infants aged 0–2 months (table 2); values reported for older children are less precise and more work is needed to assess them. The major difficulty of this method is the measurement of total daily energy expenditure. The use of the doubly labelled water method enables the measurement of TDEE of infants in a noninvasive manner [18].

The energy expended to support the metabolic cost of growth (E_m) has been obtained by regression analyses as the slope of sleeping energy expenditure (SEE) to weight gain. Consistent values of about 3 kJ/g weight gain have been reported for low-birth-weight infants in the first month of life. Thus E_g, calculated as $E_s + E_m$ is about $10 + 3 = 13$ kJ/g in these infants. For a mean weight gain of 15 g/kg·day, the total E_g amounts to 195 kJ/kg. Assuming a mean value of MEI of 450 kJ/kg·day, the cost of growth represents 43% of the energy intake in these infants.

The Importance of the Composition of the Weight Gain

Both the energy stored (E_s) and the energy expended for growth (E_m) depend upon the composition of weight gain. By using a protein concentration of 16 g/100 g in lean tissues [2], the energy requirement for weight gain theoretically varies from 9 kJ/g for a weight gain resulting only from lean tissues, to 45 kJ/g for a weight made of 100% fat. Since in infants the range of the percentage of fat in weight gain is between 10 and 50%, the range of the energy cost of weight gain can be calculated as 12.6 to 27 kJ/g, respectively [17].

In conclusion, the energy cost of growth in infants is a function of the composition of the weight gain; values between 12 and 25 kJ per g of weight gain have been calculated, these values mainly depending on the fat content of the weight gain. In low-birth-weight infants, the total cost of growth is larger than 40% of their energy intake.

Acknowledgments

The contribution of Dr. J. L. Micheli and Y. Schutz in the studies carried out in Lausanne on the energy cost of growth in infants is gratefully acknowledged.

References

1 Acheson KJ, Flatt JP, Jéquier E: Glycogen synthesis versus lipogenesis after a 500 g carbohydrate meal in man. Metabolism 1982;31:1234–1240.

2 Fomon SJ, Haschke F, Zielger EE, Nelson SE: Body composition of reference children from birth to age 10 years. Am J Clin Nutr 1982;35:1169–1175.

3 Roberts SB, Young VR: Energy costs of fat and protein deposition in the human infant. Am J Clin Nutr 1988;48:951–955.

4 Flatt JP: The biochemistry of energy expenditure; in Bray G (ed): Recent Advances in Obesity Research. London, John Libbey, 1980, vol II, pp 211–218.

5 Pencharz PB, Steffee WP, Cochran W, Scrimshaw NS, Rand WM, Young VR: Protein metabolism in human neonates: Nitrogen-balance studies, estimated obligatory losses of nitrogen and whole body turnover and nitrogen. Clin Sci Mol Med 1977;52:485–498.

6 Catzeflis C, Schutz Y, Micheli JL, Welsch C, Arnaud MJ, Jéquier E: Whole body protein synthesis and energy expenditure in very low birthweight infants. Pediatr Res 1985;19:679–687.

7 Young VR: Some basic aspects of protein-energy interrelationships; in Scrimshaw NS, Schürch B (eds): Protein-Energy Interactions. Proc IDECG Workshop, Waterville Valley, USA, 1992, pp 9–19.

8 Ashworth A: Growth rates in children recovering from protein calorie malnutrition. Br J Nutr 1969; 23:835–845.

9 Kerr D, Ashworth A, Picou D, Poulter N, Seakins A, Spady D, Wheeler E: Accelerate recovery from infant malnutrition with high calorie feeding; in Gardner L, Amacher P (eds): Endocrine Aspects of Malnutrition. Santa Ynez, The Kroc Foundation, 1973, pp 467–479.

10 Spady DW, Payne PR, Picou D, Waterlow JC: Energy balance during recovery from malnutrition. Am J Clin Nutr 1976;29:1073–1088.

11 Jackson AA, Chir B, Picou D, Reeds PJ: The energy cost of repleting tissue deficits during recovery from protein-energy malnutrition. Am J Clin Nutr 1977;30:1514–1517.

12 Stettler N, Schutz Y, Micheli JL, Jéquier E: Energetic and metabolic cost of growth in Gambian infants. Eur J Clin Nutr 1992;46:329–335.
13 Brooke OG, Alvear J, Arnold M: Energy retention, energy expenditure and growth in healthy immature infants. Pediatr Res 1979;13:215–220.
14 Reichman BL, Chessex P, Putet G, Verellen GJE, Smith JM, Heim T, Swyer PR: Partition of energy metabolism and energy cost of growth in the very low-birth-weight infant. Pediatrics 1982;69: 446–451.
15 Gudinchet F, Schutz Y, Micheli JL, Stettler E, Jéquier E: Metabolic cost of growth in very low-birth-weight infants. Pediatr Res 1982;16:1025–1030.
16 Freymond D, Schutz Y, Decombaz J, Micheli JL, Jéquier E: Energy balance, physical activity, and thermogenic effect of feeding in premature infants. Pediatr Res 1986;20:638–645.
17 Butte NF, Wong WW, Garza C: Energy cost of growth during infancy. Proc Nutr Soc 1989;48: 303–312.
18 Roberts SB, Coward WA, Schlingenseipen KH, Nohria V, Lucas A: Comparison of the doubly labeled water ($^2H_2$ ^{18}O) method with indirect calorimetry and a nutrient-balance study for simultaneous determination of energy expenditure, water intake, and metabolizable energy intake in preterm infants. Am J Clin Nutr 1986;44:315–322.

Prof. Eric Jéquier, Institut de Physiologie, 7, rue du Bugnon, CH–1005 Lausanne (Switzerland)

Subject Index